'It is one world. And it's in our care. For the first time in the history of humanity, for the first time in 500 million years, one species has the future in the palm of its hands. I just hope he realises that that is the case.'

Sir David Attenborough, *Blue Planet II*, 2017

....

PRAISE FOR THIS BOOK:

'So here are the th..r head in the beach a...e nightmare has gone......................................gardless hoping that 'they' will fix the.................note – 'they' won't, 'they' won't even try until it's ~~so~~ late) or you can get up and get on with changing the world yourself. Which means that actually there isn't a choice at all. Read this book, think and then act – it's our only hope.'

Chris Packham, Conservationist and
MCS Ocean Ambassador

'One for the dedicated eco-warriors, this book provides tips for protecting our oceans and you might also find that you save money in the process' *Independent*

'Inspiring' *In the Moment* magazine

marine conservation society

The Marine Conservation Society is the UK's leading marine charity. We grew from the hard work and forward-thinking of a number of people who, in the 1970s, could see the harm being wreaked on our marine environment. We started out as the Underwater Conservation Society and became the Marine Conservation Society in 1983, led by marine scientists with a network of dedicated volunteer supporters. Our role is to educate and inspire people to change their habits, opinions and preconceptions to help preserve our oceans for generations to come. We've treated our seas with too little respect for too long – too many fish have been taken out, too much rubbish put in. Now our seas are paying the price for the years of neglect and unrestricted damaging activities, with rising levels of pollution and species in decline. Our work has resulted in the creation of marine protected areas around the UK, plastic carrier-bag charges to reduce litter and supermarkets being more aware of the need to stock sustainable seafood. We've made sure that more people than ever know the dangers of marine litter, especially plastic litter.

You can support our work – see www.mcsuk.org.

HOW TO LIVE PLASTIC FREE

A day in the life of
a plastic detox

Marine
Conservation Society

The right ... ...nor of

Cataloguing in Publication Data is available from the British Library

ISBN 978 1 4722 5982 0
eISBN 978 1 4722 5979 0

Typeset in 10/14 pt Weiss by Jouve (UK) Milton Keynes

Printed and bound in Great Britain by Clays Ltd, Elcograf S.p.A.

Headline's policy is to use papers that are natural, renewable and recyclable
products and made from wood grown in well-managed forests and other
controlled sources. The logging and manufacturing processes are expected
to conform to the environmental regulations of the country of origin.

HEADLINE PUBLISHING GROUP
An Hachette UK Company
Carmelite House
50 Victoria Embankment
London
EC4Y 0DZ

www.headline.co.uk
www.hachette.co.uk

To the men and women who, across the world, dedicate their lives to the conservation of nature.

To the men and women of the Marine Conservation Society, who have been part of this 'extended family' for thirty-five years.

To our volunteers, who have joined us on thousands of beach cleans – they've removed millions of items of litter from our coastline, not giving them the opportunity to reach the world's oceans.

To the ocean, that never ceases to inspire us.

FOREWORD

'When I became an Ocean Ambassador for the Marine Conservation Society, I knew the charity, with their deep knowledge and understanding of the beach litter issue, was trying to intelligently tackle the topic. This book is testament to that. It offers simple straightforward advice for everyday life. If you're trying to make changes at home, this is a brilliant handbook written by people who, like you and me, live "normal" lives and who have witnessed the struggle, first-hand, when trying to live a plastic-free life. A good read from cover to cover or a pick-up-and-put-down book, it's full of advice on going plastic free from the time you get up until the time you go to bed with all the activities you can think about in between – babies, holidays, pets, cooking, clothes – they've left no stone unturned. And all with a good dose of humour and history thrown in! I love the sea and the coast and have always been passionate about the health of the ocean, so when it comes to taking on board the tips in this book – I'm in.'

Deborah Meaden, Entrepreneur, BBC *Dragon's Den* investor and MCS Ocean Ambassador

CONTENTS

OTHER IMPORTANT TIMES TO DETOX:

WHEN NOT TO DETOX:

INTRODUCTION

IT'S TIME FOR A PLASTIC DETOX

This is not a book against plastic. It is a journey through a typical day in our lives. One made with fresh eyes.

We're so used to it we don't realize that, from the moment we open our eyes in the morning, most of what we see is made of plastic. And we're so used to it that we've stopped asking the important questions: is this necessary? Where does this plastic come from? Where will it end up once it has done its job?

This book will help you 'treat' your (often unrecognized) plastic addiction. It will make you notice things you're so used to that you've stopped minding them: why is there a straw in my drink, why a stirrer? Why is my sandwich wrapped in plastic, and my coffee, my fork, my salad, my toothbrush and toothpaste, the shampoo, the kids' toys? My clothes. My chair and table, the carpet and the ceiling. How did that happen? How do we change it?

THE ENVIRONMENTAL EMERGENCY

Choking, starving, poisoning. That's what plastic litter is doing to marine life and as our understanding increases, so does the horror. Not a day goes by without new scientific evidence making the headlines about the plastic plague that is enveloping our seas, like a virus with no known cure.

At first it was the stranded, starved animals with stomachs full of plastic that told us how lethal the light, sturdy and practical material we use everyday is, once out at sea. Then, with research progressing, the plot thickened and became even gloomier. Plastics in the ocean are not just mistaken for food and ingested, they're creating a toxic nightmare that's messing with Mother Nature. We know now that man-made chemicals cling to plastics, like the crew of a sunken ship clinging to a life raft. With an estimated 300 billion pieces of plastic floating in the Arctic Ocean alone,[1] all acting like a toxic sponge to carcinogenic and endocrine-disrupting chemicals (substances that cause cancer and serious dysfunctions in our hormonal balance) such as PCBs, BPAs and pesticides, our seas are facing horrors on an unprecedented scale.

Animals that have survived for centuries in the remotest parts of the world are now playing a role in a real-life drama – their bodies are being changed by an unseen enemy. Polar bears, for example, are not just up against climate change. Chemical pollutants have been found to weaken their immune system and are playing havoc with their reproductive cycle. Toxins are also disrupting the hormones of both males and females.[2]

A mating disaster. Seal pups, irresistibly cute with their big, sad eyes, are being born already affected by high levels of toxins like PCBs. The pups have reduced immunity due to the transfer of these chemicals from their mums, making them more vulnerable to infection. These toxins have also negatively impacted seal reproduction.[3] Hideously, spontaneous abortions in Californian sea lions have also been linked to these chemicals.[4]

Albatrosses form lifelong partnerships, and after roaming the open seas in solitude they return to the same mate, in the same place, year after year, to raise their chicks. Pretty amazing. But for Sir David Attenborough they're the protagonists of one of the most heartbreaking stories of *Blue Planet II*. He shared with all of us a moment in which he witnessed an albatross mum and dad, returning from their long and arduous fishing expeditions, only to unload not sand eels, fish or squid, but . . . plastic.

And the gallery of horrors continues. Billions of nurdles – the plastic pellets that are used in the plastic industry – are lost into the environment annually. They're the same size and shape as fish eggs, and, sadly, many species of seabird feed this toxic plastic to their chicks. Like the albatross, our very own fulmar population is also affected, with plastics, including nurdles, being found in the digestive system of over 90% of tested birds.[5]

In 2016 the Ellen MacArthur Foundation produced a controversial report with some truly disturbing conclusions: the equivalent of one entire truck of plastic (8 tonnes) is dumped in the sea every minute and, in a business-as-usual scenario, by 2050 we might have more plastic than fish (by weight) in the sea.[6] You don't need to be a passionate diver or snorkeller to dread the horror of that moment.

THE TIME HAS COME TO CHANGE ALL THIS

We must change the way we produce and consume plastic. The solutions are readily available and economically viable. None will require us to adopt monastic lifestyles. These solutions will get increasingly cheaper, and innovation will create more answers, as the market for nature-friendly products increases. And this should, in turn, provide a genuine contribution to the global issue as often it's these very innovations that can be adopted across the world to address environmental issues. Examples abound: solar panels, windmills and mobile technology, for instance.

BUT WHAT CAN WE DO, AS INDIVIDUALS?

To start with, we can fix our own littering habits. Our yearly stats shockingly revealed in 2017 that plastic litter on UK beaches is on the rise. We're using the great outdoors as a huge dustbin and much of our 'on the go' litter is ending up in the sea and on our beaches. In September every year, we run The Great British Beach Clean, the UK's largest and most influential national clean-up and survey of Britain's beaches. The 2017 event saw 6,944 volunteer beach cleaners pick up record amounts of litter over four days from 339 UK beaches – a staggering average of 718 bits of rubbish from every 100 metres cleaned. It's a 10% rise in the amount of beach litter picked up compared to 2016. Tiny bits of plastic and polystyrene were once again top of our list of finds in 2017, but there are many others that we have a real opportunity to address. We are talking about coffee cups, plastic cutlery, plastic wrappers, straws, sandwich packets, lolly sticks, plastic bottles, drinks cans, glass bottles, plastic cups, lids and stirrers. 138 pieces of this specific type of litter were found, on average, per 100 metres of all the beaches cleaned and surveyed by our volunteers.[7] We go through some of these ugly finds in more detail in Chapter 9.

FROM A THROWAWAY SOCIETY TO THE CIRCULAR ECONOMY

But it's not just about littering, or the impact that this has on wildlife. When the MCS President, HRH the Prince of Wales, told the Our Ocean summit in October 2017 that the world had reached a critical point where plastics are 'now on the menu', he probably raised quite a few eyebrows. Prince Charles was referring to the fact that plastic is increasingly found in fish.[8] We are yet to fully understand the impact this will have on our health. But it doesn't look too promising.

To reduce the amount of litter that's in our seas we need to take a long, hard look at the way we've all chosen to live. Our world has been filled with cheap things that we don't value. We're leaving the contents of our lives behind on our travels, or we're flushing it down the loo, or stuffing it in the bin and letting it go off to landfill. Much of this life waste is ending up in our seas and because so much of it is made of plastic we're lumbering our oceans with a massive, massive problem.

We're constantly throwing things away nowadays; we leave anything and everything behind. Can you imagine anyone going camping in the 50s and 60s and leaving their tent behind? Go to any summer festival now, from Glastonbury to Creamfields, and you can't move for abandoned tents when everyone's gone home. (See Chapter 6 for some ideas on how to change these habits.)

Dealing with waste is a rather unprecedented issue in the history of our planet. The fact of the matter is that there is no such thing as waste in nature. Everything, no matter how small or big, smelly or dirty, is reused by some other organism in one big circle of life. The law of nature is simple: everyone's debris becomes someone else's raw material. That was until we invented, just a few decades ago, materials that would take centuries to decompose and that would, in the meantime, be beneficial to few and lethal to many species. We humans invented damaging and poisonous waste. Waste not just unprecedented in persistence but also unimaginable in quantity.

It is not too late to change things. Throughout this book you'll find lots of tips for reducing your plastic consumption and your production of waste.

We simply need to replicate, in our contemporary lives full of gadgets and lasting materials, the way nature works. How? Imagine a world where everything is designed to either last, or to become something else (recycled in its entirety or in its parts/materials) once its lifetime is over, its mission accomplished. The transition from a throwaway society to a circular economy will require radical changes in the way we produce and consume. And the way our governments here in the UK use our taxes will play a huge role. But, just like a Chinese philosopher once said, a journey of a thousand miles begins with a simple step.

The transition to a new, cleaner world starts today with us finally remembering to tell the bartender 'I do not want a straw in my drink, thank you'. The time has come to stop the plastic tide.

WHAT CAN YOU DO?

You'll be surprised at what you can do to make changes. You don't have to completely ditch plastic from your life. Some of the tips in this book are about going back to basics, doing what people have done for thousands of years, with no inkling of the plastic era to come. Other things you can do are very modern solutions, using new techniques and technologies to leave plastic behind us.

We all like to buy new things, and find new ways to enjoy spending our leisure time. Most of us aren't deliberately wasteful, thoughtlessly uncaring about the impacts of our rampant consumerism on the planet and on our children's future. It's just that it can be difficult to find the necessities (and occasional indulgences) of daily life without a plastic container, wrapper or freebie throwaway toy.

We've looked long and hard for the best ways to ditch plastic that we use in everyday life for this book. Sometimes there is no simple answer. For example, how do you keep food fresh? There seems to be a thin wrapping on almost everything sold in the supermarket, in any high street or out-of-town outlet, and even in the local corner shop, bakery or butcher. This book will help you look at things you're used to in a new way.

You'll find notes pages at the end of every chapter for you to keep track of useful websites, sustainable shops and any other tips that you discover on your plastic-free journey.

THIS IS NOT A BOOK AGAINST PLASTIC

Plastic is a phenomenal material; it is light, durable and sturdy – and it can save lives. We've dedicated our last chapter to some of its amazing uses (hospitals, for example, could not be run without plastic nowadays).

The problem is, firstly, that a lot of plastics today are neither recycled nor recyclable, and so we keep adding mountains of plastic litter to an already intoxicated planet. And secondly, most of the plastic we're surrounded by is simply not necessary.

The world we need to transition to is one where we use only (or mostly) recycled plastic that is recyclable. A world in which we ditch all useless plastic (not only bags, straws and cutlery but also toothbrushes, razors, etc.) and we give good plastic a value, so that it won't be buried in the ground.

This book will, hopefully, help you do just that.

On behalf of the Marine Conservation Society (and all sea creatures) . . . thank you for trying.

Clare, Luca and Richard

'In 2015, UK householders threw away over 663 tonnes of electric and electronic gadgets, over 10kg per person. The worst part? Almost a quarter of it may still work.'

1

GETTING UP

07:00

Most of us struggle with that first moment of the day: the alarm going off.

Some of us might hit the snooze button with excessive force, as if it were the clock's fault that we have to leave the warmth and comfort of the bed. Some might even hit the clock so hard they actually break it.

And some, looking at the lifeless object on the floor, might have thought: I can repair it. There was a time, not so long ago, when this was a perfectly reasonable idea. A (typically male) teenager in the 80s for example, might have owned one of the first PCs: a Sinclair or a Commodore. Back in those days, news stands were full of magazines on electronics and teenagers could actually design and build circuits to add 'powers' to their small machines. Everything was possible, we owned and controlled the machines in our life.

But trying to fix a contemporary (inevitably made of plastic) alarm clock is a different business: it is almost impossible to disassemble, almost as if it's designed that way. And if you persist, it will probably break apart. Laying out its components will reveal a micro world of parts made with the minimum possible amount of costly materials (metals, electronic components).

The plastic alarm clock, the simple everyday object, is a great example of our race towards increasingly short-lived, cheap electrical goods.

It's not all negative, of course. If you were a teenager in the 80s your PC had 16 KB of memory. Thirty years later the PCs like the one on which this book was written have a memory 100 million times greater. And, despite the fact that most of us are still doing the same thing with a PC (i.e. writing), you'd be wrong in thinking not much has changed, because contemporary electronics have changed people's lives for good.

THE END OF THE WATCHMAKER

Some of us will still remember the old alarm clock: round and made of steel, with two small bells on top and the little hammer in the middle. There was no need for batteries, you could charge it every night with a little key, and a spring would keep it going (and cause it to ring furiously in the morning). Today, people in their twenties have probably never seen one. Let alone kids.

When 'that' kind of clock broke, we could take it to one of the thousands of watchmakers in the UK, who were among the army of 'Mr Fix-its': ingenious individuals running shops that contained a huge amount of spare parts and old machinery. These people could fix pretty much anything: TVs, radios, washing machines. Most of these jobs don't exist anymore, because they're not needed. Today a brand-new alarm clock only costs a few pounds and fixing it, if one could still find Mr Fix-it, would probably cost a lot more. Initially workers moved into production or industrial services (sales, marketing, etc.) but then, with globalization, factories moved to countries where wages were lower. Over the last few decades, the UK has lost half of its jobs in manufacturing – from 5.7 million in 1980 down to a mere 2.7 million in December 2017.[9]

The mass production of disposable goods has not only destroyed a lot of jobs in industrialized countries, but it has created a huge abundance of waste – of all kinds and in impossible quantities.

A CONTEMPORARY WASTELAND

The plastic alarm clock, in pieces, is now electronic waste (e-waste) and not easy to dispose of in the right way (i.e. making sure it is recycled or at least that it doesn't pollute and contaminate). According to a UN study, only 20% of e-waste, worldwide, was recycled through appropriate channels in 2016.[10] The remaining 80%? Probably in a hole in the ground.

And the amount of waste is not the only thing to worry about; the plastic, along with other components, contains a number of rather toxic chemicals and heavy metals that, should they leach into the ground and into our water supply, would be very harmful to our health, and to the health of our wildlife.

However, because of strict regulations on environmental, health and safety procedures, recycling e-waste in the UK comes at a cost. So the e-junk finds its way back to less developed countries (such as China, Ghana, Bangladesh) where it's often disassembled and recycled with few, if any, precautionary measures, by underpaid (and often under-age) workers. The metals in the parts will be sold back into the market and the plastic will either end up in a landfill, be burned (releasing yet more toxic chemicals) or turned into a chew toy for our children and grandchildren.

Anecdotes regarding the illegal trade and dumping of e-waste abound. Some old junk PCs, for example, are falsely declared as second-hand goods or educational material for charitable projects for schools in Africa. A few years ago the Environmental Investigation Agency placed a GPS inside an old PC from the UK and traced it all the way to a pile of rubbish in Ghana.[11]

The UK is a world leader in the production of this high-tech 'junk'. Interestingly, a couple of years back, UK charity Wrap published a study that revealed 23% of it was actually functioning fine and only needed minor repairs.[12]

The short somewhat simplistic version of this chapter could be: when you buy a plastic alarm clock made in China you are contributing to the following chain of events: 1) someone could lose their job in the UK 2) soon you'll produce an e-waste that 3) will go back to some developing country to be recycled by underpaid workers in unhealthy conditions and 4) parts of that junk will find their way back to your home. It could become a plastic toy or another cheap electronic gadget.

This vicious cycle goes on and on. All for a few pounds.

5 THINGS YOU CAN DO TODAY TO DETOX YOUR WAKE UP FROM PLASTIC

1. Honestly, you do not need an alarm clock. All contemporary phones have an alarm function. So don't buy one if you don't have one.

2. If you do have an alarm clock, don't upgrade it to a newer model if it's still working fine.

3. When your alarm breaks, if it cannot be fixed, go online and find the nearest recycling station. DO NOT, in any circumstance, throw your old alarm clock/PC/radio in the bin.

4. If you really want one, go for an 'eco-friendly' alarm clock. Make the first thing you see when you wake up a reminder of your commitment to sustainability. Choose one made from renewable materials, or a beautiful vintage one that doesn't use batteries.

5. In general, buy higher-quality goods. They may be a bit more expensive but they'll last.

NOTES

....

....

....

....

....

....

....

....

....

....

....

....

....

'Between 2016 and 2017 the Marine Conservation Society Great British Beach Clean revealed the number of wet wipes found on UK beaches had doubled. In a single year!'

2

THE BATHROOM

07:15

For many people their favourite room in the house isn't the
kitchen, the bedroom or the sitting room; it is, in fact, the
bathroom. A room with clean white lines, from which you
emerge all sparkly and ready to face the world, scrubbed up
and with your face fully on. It's also often the only room in
the house where you can legitimately get away from it all,
somewhere hygienic, private and comfy.

A QUICK HISTORY

Water is essential to our survival and, for this reason, early
human settlements were all near a source of fresh water. But
when it comes to actual bathing, our relationship with water
has been tortuous. In *Clean and Decent: The Fascinating History of
the Bathroom and Wc*, Lawrence Wright wrote: 'The bath was
brief, cold and invigorating in classical Greece. In Rome and

Islam it meant relaxation, bodily refreshment and well-being'. All variations of this rite have been explored in the past 4,000 years in a totally non-linear fashion. In 1960 Wright could say:

'A living Englishman has complained of his Oxford college that it denied him the everyday sanitary conveniences of Minoan Crete [about 2,000 BC]. The 15th-century gentleman used the bath, but his 17th-century descendant did not. The monk of 1350 enjoyed more orderly plumbing, and sweeter habits, than the Londoner of 1850. The Polynesian "savage" was cleaner than either.'[13]

Throughout history, there has been a religious element to bathing – people cleansed themselves before entering a sacred area. But it was the Romans who took bathing to a whole new level. They were obsessed with hot baths or *thermae*. They loved a straight road but they loved a big hot bath even more, and they treated bathing as a community activity.

Once the Romans had conquered new lands they couldn't wait to build thermal baths. Posh Romans had their own baths at home, of course, but rich or poor could use these large public facilities. Romans would spend hours in there just floating and scrubbing – and presumably shrivelling up, too. Politics were discussed, business was done, gossip was spread. They'd lie in a hot pool, plunge into a cold one, and then repeat the whole process.

The popularity of bathing took a bit of a plunge itself in the Middle Ages, when water was seen as a carrier of disease and a hefty douse of perfume was preferred. Interestingly, many people started making their own soap at home during this time! Nevertheless, in the 14th century, it was thought the plague was spread by disease entering through the pores; because bathing opened the pores, people were advised against it. Keeping dirty and layered up with filth was thought to keep the disease at bay. Sounds bonkers now, but as one in three people were catching the plague, and most of them dying from it, you can see the Middle-Age logic. Spring forward through the 16th, 17th and 18th centuries and public bathing declined, with bathhouses shut down in favour of bathing at home.

Copper baths replaced the old wooden ones and were followed by cast iron lined with porcelain and finished off with claw feet. By the 1920s, council houses in Britain were built with bathrooms, but even so, as late as the 1960s, many homes in Britain didn't have a bathroom, which meant that some people still had an outside loo and had to make do with a stand-up wash or a dunk in a bath that was placed in front of the fireplace. The man of the house, however filthy, would always take a dip first and the baby would go last. Was a baby ever thrown out with the bathwater? We'll probably never know!

THE MODERN BATHROOM: IT'S JUST YOU AND A SHEDLOAD OF PLASTIC

With so many people embracing the concept of going plastic free by turning their backs on plastic bottles, straws, carrier bags and coffee cups, many have ignored the fact that things have gone very badly wrong in the smallest room – the bathroom.

This haven of heat and steam, this palace of peace and potions is so full of plastic that it's perhaps the most difficult place to wage war on. To start with, most bathrooms have a lot of plastic hardware in them: the shower tray and screen, the bath itself and the toilet seat – whether it's up or down! However, if you're thinking big in a bid to de-plasticize your bathroom there are plenty of ways to do it.

The bath

You can go all fancy and fork out some proper cash for a steel bath coated in enamel instead of the ubiquitous acrylic bath (acrylic is a type of plastic). Although pricey, steel and enamel tubs look stunning. Run your hand over them and they shriek 'quality', 'decadence', 'shove your cheap plastic alternative'. The brilliant thing about these baths is that when you decide yours has had its day, it can be safely recycled because it's made of steel.

The shower tray

Shower trays come in two basic materials – acrylic or stone resin that mimics natural stone. Acrylic can move and can therefore leak, whilst stone resin is harder wearing and cleans up better. 'You pays your money and takes your choice', as they say . . . but please don't make plastic your shower tray of choice.

Of course, if you do have an acrylic bath or shower tray and they're currently working fine, there's no need to create unnecessary landfill by removing them.

The loo

Give your rear end a nice treat with a natural wood seat.
Warmer to the touch, these seats give your bathroom a lovely
traditional feel. The good news is that the toilet bowl itself is
made from something called vitreous china. This is brilliant
stuff that makes the loo waterproof and stainproof through its
entire thickness. Your pan is not plastic – what a relief!

POTIONS AND LOTIONS, WIPES AND GELS

Bathrooms are brimming with them but pampering doesn't
come plastic free.

Shower gels and exfoliating washes

Like the Romans, the Ancient Egyptians before them liked to
feel clean. They used pastes made of ash or clay and perfumed
with oils and plant extracts such as, henna, cinnamon,
turpentine, iris, lilies, roses and bitter almonds, which created
a lather on the skin, or they mixed animal and vegetable oils
with alkaline salts to form a scrub. All very wholesome and all
very natural and biodegradable.

Fast-forward over centuries and the watchword for cleanliness
amongst the Victorians was 'carbolic' – a rough soap
containing the disinfectant phenol or carbolic acid. Whilst
some women cracked eggs on their heads to work up a natural
lather, others were using lye, a very caustic chemical. Our
ancestors were changing – from being lovers of the natural to
going down a more chemical route.

Then we come to the modern day. How many shower gels have been sold to us which highlight their exfoliating properties, their label telling us they'll leave skin revitalized and reinvigorated? Sadly, many contained (and some products used in the bathroom and around the house may still contain) Microbeads – and there's nothing natural about those.

Patents for plastic Microbeads in personal-care products were introduced in the 1960s. Natural went out of the window as they were popped into lotions, face washes, toothpastes, shampoos and exfoliators. They didn't sound too bad, these little beads. Surely they must be harmless if they're so tiny? Tiny they may be, exfoliating too, but harmless? Wrong.

Microbeads are solid plastic particles of less than one millimetre wide. They're usually made of polyethylene but can be polypropylene or polystyrene. They've been sold for decades without us really realizing, and washed down the sink. So tiny, they can't be filtered out by wastewater treatment, and many tonnes of these little blighters have flowed into rivers and seas every year; they harm wildlife and can even end up in our seafood.

But there's good news. In the UK, manufacturers are now banned from adding them to rinse-off cosmetics and personal-care products such as face scrubs, toothpastes and shower gels. This is the result of a campaign by the Marine Conservation Society and other partners (see Acknowledgements), which we're really proud of.

Microbeads are still used in cleaning products, like dishwasher tablets and washing machine sachets, so are not gone altogether, and you do need to be careful not to inadvertently buy cosmetic products containing them when overseas.

There are plenty of natural exfoliants we can use instead of Microbeads: crushed cocoa beans, ground almonds, ground apricot pits, sea salt, ground pumice and oatmeal – even your used coffee grounds!

Soap

You don't have to go very far back, the 1960s will do it, to find the bathroom era of bars of soap. Families had their go-to favourites like Lux toilet soap (an odd name seeing as these bars never actually went down the toilet). Back then soaps were slippery, didn't lather up too well and, because there were fewer showers in our bathrooms, they just reclined in the little dimple on the basin or the bath. But soap is back and it's better than ever – and it's plastic free.

A soap has finally been invented that won't fly out of your hands. It's designed like those big concrete tetrapods that are used as breakwaters to stop coastal erosion – it's tetra soap! Seriously, it's been done: tetrasoap.com. It's organic and sweet smelling, it produces a rich lather and is designed to last. You can also now buy bars of soap without any sort of plastic packaging from high-street outlets and on the internet.

Hair care

Like shower gels, the nation's favourite haircare products, in the main, come in plastic containers that end up in the bin once the product has run out. The answer to a plastic-free alternative for shiny locks also comes in the shape of a bar. Shampoo bars can last three times as long as a regular bottle. Head to the Lush website (lush.com) – they've got heaps of shampoo bars with lots of testimonials. But a quick internet search will throw up a whole load of alternatives. You'll need to hunt around as some may dry your hair out.

Switching to a shampoo bar may be a more experimental journey . . . but it will be totally worth it in the end. You can even get wooden combs and brushes for the hair you're now washing with a shampoo bar.

Keeping fresh

Making your own deodorant is a great alternative to a plastic-packaged one. Simply mix coconut oil and bicarbonate of soda with one drop of anti-bacterial essential oil and you're away!

Of course, if you're on the sweatier side you may feel a bit safer buying something rather than whipping up your own in the kitchen. If this is the case, there are plenty of natural sticks and bars about. You just need to check the packaging as some do come in a bit of plastic.

Toothbrushes and pastes

Teeth-cleaning implements have apparently been around since 3000 BC, when our ancestors used a 'chew stick', a thin twig with a frayed end which they rubbed against their teeth and gums. It was the Chinese in the 1400s who invented the bristle toothbrush, attaching hogs' bristles to handles made from bone or bamboo. Hogs' hair gave way to horsehair, as we Europeans liked a softer brush, then pig bristles and badger hair were used. Nylon bristles replaced animal hair in the 1930s and now the handles are made from polypropylene and polyethylene – better known as plastic. All very convenient to fend off cavities but very inconvenient for the environment.

According to the dental profession, we should replace our toothbrush every three months, thereby adding to the growing plastic mountain. But there's good news on the toothbrush front. The bamboo brush is back! Although it takes some getting used to, it's fully biodegradable and once you've got past the less-than-familiar feeling in your mouth you're good to go! There are a number of other 'green' brushes on the market, including ones made from yoghurt pots if you prefer a more familiar plastic-y feel.

Most toothpastes are packaged in plastic, whether they're in a tube or a pump-action canister. Swap your traditional toothpaste for toothpowder, which you can buy on the high street or online – just make sure you get one in a tin that you can then reuse, perhaps for your own home-made version!

Home-made plastic-free toothpowder

4 tablespoons organic coconut oil

2 tablespoons baking soda

1 tablespoon bentonite clay (food grade)

20 drops peppermint oil (food grade or essential oil)

1. Mix all the ingredients together until you have a paste and pop into an airtight glass jar or small tin.

2. Before using, wet your toothbrush, shake off any surplus water, press into the powder then brush your teeth in the normal way.

Before fluoride was introduced into toothpaste, people brushed their teeth with baking soda because it's a mild abrasive that both polishes and whitens. Its antibacterial activity also kills bacteria that cause tooth decay. Bentonite clay contains calcium and binds to the bad bacteria around your teeth, tongue and gums to help remove them before you spit it out.

Home-made plastic-free mouthwash

Makes 1 batch

½ cup of filtered water

2 teaspoons baking soda

2 drops tea tree essential oil

2 drops peppermint essential oil

Add everything to an airtight glass jar, give it a good shake then swill around your mouth for about a minute before spitting out.

Store in the fridge for up to a week.

Make-up

If you're going plastic free then make-up can be a bit of a minefield. So many plastic containers and pots, it's hard to know where to begin. But plastic-free make-up does exist and a quick search of the internet will be a real eye opener! There are plenty of brands out there which will turn you from a plastic pamperer into a natural woman without having to bare all. Try RMS Beauty (rmsbeauty.com), Kjaer Weis (kjaerweis.com) or Fat and the Moon (fatandthemoon.com) but there are other options, too.

Make your own

Did you know you can make your own mascara? Spend a bit of time doing some experimental mixing with ingredients like coconut oil, charcoal, beeswax and finely grated soap and you could find the perfect plastic-free alternative. There are plenty of recipes and methods on the internet and, to be honest, it's a bit of fun!

Make-up that's gluten free, cruelty free, vegan, vegetarian, totally without toxins, sustainably packaged and even refillable . . . it's all out there if you care to look and lots of it is free from plastic. Plastic-free lipstick, foundation, mascara, eyeliner, blusher and bronzer – you'll find it somewhere on the internet and if you spend a few hours searching, you'll come across some good blogs and reviews that will tell you how everyone who's like-minded to go plastic free has got on with it.

Choosing your make-up from the internet can involve a lot of trial, error and guesswork – it may be plastic free and totally natural, but will it be the right colour when it arrives? The answer is to start your plastic-free beauty journey with a friend. Ideally someone that uses a similar palette of colours to you or has similar skin tone. By working as a team you can buy a few different shades and types – creams, powders, water-based foundations – and then swap if your shade isn't exactly bang on.

Make-up removal

So, once you've nailed your plastic-free beauty regime, you will need to get it all off at the end of the day. Up until now this is where the make-up wipes have swung into action. Their sales growth over the last decade has been exponential. Cheap and simple to use – simply chuck them in the bin when you've finished – what's not to like? Well, pretty much everything, really.

Make-up wipes are usually labelled as 'non-flushable', although that is not always the case. Their close relatives, wet wipes, are found in their thousands on the beaches of the UK. Between 2016 and 2017 the MCS Great British Beach Clean revealed a 94% rise in the number of wet wipes found on UK beaches. In just a single year! They get there because labelling is at best confusing and at worst just plain wrong. If it says 'flushable', well, you'll flush it won't you? But make-up wipes, along with other household wipes, won't break down and they contain plastic – although thankfully many manufacturers are removing the plastic content of wipes and retailers are making the labelling a lot clearer so they're not flushed down the loo.

But how else do you remove make-up? Simple – go back to basics. A flannel will do the trick, or a little coconut oil and a cotton pad. Or you could go all exotic and use a konjac sponge made from the konjac potato plant found in Asia at very high altitudes.

Having a plastic-free bathroom regime is a bit more expensive, quite a bit more in some cases. But anecdotal blogs and reviews of many of the products reveal that a lot of them last longer than their plastic relatives and, of course, you can't put a price on securing the long-term future of our oceans.

MENSTRUAL MANAGEMENT

Periods – still not a subject that anyone seems terribly comfortable chatting about openly, despite them being around for, well, forever.

Not only are they an inconvenience for women – does anyone ever look forward to their next one? – but managing them is also contributing to our plastic crisis. During the MCS Great British Beach Clean in 2017 there was yet another increase in sewage-related debris on British beaches and that included hundreds of menstrual pads, tampons and applicators.

Until disposable sanitary pads were created, the first contraptions were all loops and belts to hold cloth or reusable pads in place. In some cultures, menstrual huts are the place women go, out of sight, to have their 'menses'. But in most industrialized countries nowadays, to make the best of a bad job, even periods have become a slave to plastic convenience.

In the UK, every woman uses an average of over 11,000 disposable menstrual products in her reproductive lifetime. Tampons, pads and pantyliners generate more than 200,000 tonnes of waste per year, and the majority contain plastic – pads are made up of around 90% plastic.[14]

But there are alternative feminine hygiene products, ones which will create less impact on the environment – particularly our oceans and beaches.

- Give a menstrual cup a go. These are made of silicone and can hold up to three times as much fluid as a tampon can absorb.

- Try reusable pads or biodegradable cotton tampons. Again, switching may require a little investigation, but converts say they are more comfortable, more hygienic and healthier.

- You could try Thinx (shethinx.com) – claimed to be period-proof underwear that can replace tampons and pads on light to medium days.

An internet search will throw up heaps of ideas and reviews to help you make monthly menstruation less polluting.

The bottom line on loo roll

You can get 100% recycled loo roll but all too often it comes wrapped in plastic which sort of defeats the point. There are options on the internet which are totally recycled and come in cardboard boxes, or search for a brand whose packaging is eco-friendly, too.

5 THINGS YOU CAN DO TODAY TO DETOX YOUR BATHROOM FROM PLASTIC

1. Replace bottles of shower gels, shampoos, and canisters of deodorant, with solid bars instead.

2. Replace your plastic toothbrush with a bamboo alternative and have a go at making your own tooth powder and mouth wash (see pages 28 and 29).

3. Stop using make-up wipes that contain plastic and NEVER flush wipes down the loo.

4. Get on the internet and look for plastic-free make-up – you'll be surprised at the choice and the packaging looks good on the shelf, too.

5. Buy recycled loo roll. Ecoleaf is made from 100% recycled paper and the packaging is said to be 100% compostable.

NOTES

....

....

....

....

....

....

....

....

....

....

....

....

....

....

....

'How can someone so small need so much stuff? Toys, nappies, bottles, strollers, cribs: babies, welcome to plastic-land.'

3

BABIES AND KIDS

07:30

You're awake, showered and perfumed. For some, it's now time to
go and get the kids ready. If you're struggling to open the door to
their room, it's probably because there's a pile of toys in the way.
Probably a pile of, you've guessed it . . . plastic.

For nine months a baby is cocooned in a plastic-free
environment, all cosy and warm. Their first encounter with
the outside world, after a cuddle, is a plastic crib. Hospital
cribs for newborns are made from an anatomical transparent
thermo-moulded plastic. Of course, it's for a very good
reason . . . the crib is hygienic and easy to clean and you can
see your baby through it. What many of us don't realize is that
it marks the beginning of a plastic-filled childhood.

Plastics can contain a number of chemicals that can seriously
affect a child. Two things make children particularly

susceptible to the exposure of chemicals in plastics: firstly, they tend to put everything in their mouths. From bottles to toys, it's all made of plastic and since some of the chemicals leach from the plastic, our children risk ingesting them. Secondly, children's organs and nervous system are fragile and undeveloped, so the absorption of chemicals can be more harmful than for an adult.

A substantial amount of research now indicates that several substances disrupt the hormonal balance of babies.[15] This leads to a vast array of potential health issues. Some notorious chemicals are phthalates. These chemicals make plastic bendy, soft and squidgy . . . all the qualities you want in a toy. Which is why you'll find phthalates pretty much everywhere. Have you ever noticed that some toys become harder or less flexible with age? That's also because, adding to the effects of heat and light, your child has sucked on the toy and ingested these softening chemicals. Phthalates have been shown to impact:

X birth outcomes, including gestational age and birth weight

X fertility (lower sperm production)

X anatomical abnormalities related to the male genitalia

Ongoing studies are now looking at the relationship between phthalates and asthma,[16] and there are also studies examining whether phthalates influence the timing of puberty and the risk of childhood obesity.[17] Because of these risks, phthalates and other substances are now banned from baby products but the reality is that we keep finding them in cheap imported toys (typically from China) and that they are in so many other products (PVC in food packaging, for example) that it is almost impossible to really live phthalates-free.

If you're having any doubts about the importance of the fight for less plastic in our lives . . . think again. Because when it comes to children this is a fight for their health. And their health is compromised by the fact that their lives have been turned into a journey through plastic-land. A journey that starts even before they are born.

PLASTIC-FREE PREGNANCY TESTING – IS IT POSSIBLE?

Pregnancy test kits are made of a fibre strip inside a plastic case. You pee on the end, the chemicals on the strip react and give you the positive or negative result. You can't reuse them. So every time you buy one and throw it away it ends up in landfill. Every time, you're adding to the tide of plastic waste. But how else will you know? You might think that before the invention of the plastic test kit it was a case of wait and see. Wrong: women have wanted 'to know' for thousands of years.

In Ancient Egypt, women peed on wheat and barley seeds over a few days. If the barley sprouted it was a boy, if the wheat sprouted it was a girl and if nothing happened then, well, nothing happened. This was put to the test scientifically in the 1960s and apparently it worked 70% of the time! In the 15th century, women were encouraged to pee in a basin, put a door latch in the pee, leave it for a bit and then remove the latch. If the latch left an imprint on the basin then pregnancy was confirmed. Handy – if you had a spare latch lying around not attached to a door. In 16th-century Europe you could train as a 'piss prophet': experts who could, apparently, tell if a woman was with child just by a cursory glance at her wee. A French doctor in the 1830s spotted that female genitalia changed colour if a woman was pregnant. This is actually generally the case. But it never caught on as doctors of the day didn't really like doing much down there . . . certainly not just having a gawp.

If you're a regular tester and you haven't got wheat and barley seeds or a latch to hand, then there's good news. A test kit claiming to be 99% accurate, 0% plastic is set to hit the market very soon. It flushes like 3-ply loo paper.

FIRST BREATH – LESS PLASTIC WASTE IN THE BIRTHING SPACE

If your baby will be born in a conventional hospital then you'll just have to accept the plastic crib and all its corollaries. Some mums, however, choose to have an old-style traditional birth at home. Just to clarify, we are not recommending that you have a home birth just to reduce the use of plastic. Always put yours and the baby's health first when deciding and consult your doctor. Having a baby at home can offer more control over plastic use compared to a hospital, where single-use plastics are everywhere. However, do not be mistaken and think that the only pieces of equipment you need to give birth at home are plenty of towels, some buckets and lots of hot water. That's in the movies. In reality, since it's a rather messy business, midwives will still need a lot of plastic sheets and some kit.

At home you have your own food on tap and won't need to rely on the hospital café. You could ask a partner or friend to make you some plastic-free high-energy snacks to keep everyone going (not least yourself) during labour. The NHS got through more than half a billion disposable cups between 2013 and 2018, according to data collected by the Press Association via a Freedom of Information request.[18] At home you can keep hydrated with endless cups of tea from your favourite mug.

However, most women will give birth in a hospital where there's plenty of plastic about. In this case your best option is to check with the hospital to see what their policies are around bringing your own drinks and snacks, glass straws, etc. to reduce the use of single-use plastic during your stay.

NAPPY-ER TIMES AHEAD

Babies emit things from both ends on an ongoing basis. Whilst the top end can be wiped up with reusable cloths, the bottom end is a bit more tricky. Three billion disposable nappies are sold, and discarded, in the UK every year. The average baby gets through about 4,000 between birth and finally being dry at night. That's about £400 a year parents spend on nappies and nappy sacks according to the Environment Agency/*Which?*.[19] And the plastics in the nappies are estimated to remain in the environment for hundreds of years. The good news is that nowadays there are biodegradable alternatives to plastic nappies. The bad news is that they come at a price and unfortunately they'll still end up in the same place: landfill. Being contaminated by the baby's poo means they cannot be added to the compost pile.

So do you ditch the convenience of the disposable nappy and take up with Terry? The environmental superiority of a cotton nappy is actually not as obvious as it might seem. A report for the Environment Agency in 2008, for example, compared the environmental impact of both types of nappy over a two-and-a-half-year period. It found that using disposables creates around 550kg of carbon emissions, whereas reusables could create up to 570kg of carbon emissions when you account for the washing (water, soap, energy).[20]

However, since those figures were published things have moved on a bit and you can buy a variety of nappy systems which are washable and reusable, have non-toxic inners and outers and are made from a variety of materials, such as bamboo and hemp. They also fit a lot better than a towel and a safety pin!

TODDLER TERRORS

If you thought babies came with a lot of plastic paraphernalia then wait until they become toddlers. Toddlers take plastic use to a whole new level – and then some. You'll know you're in a house with a toddler because every surface is covered in plastic. No one has yet been able to answer the fundamental question: how can someone who is only about three feet tall require so much stuff?

When a baby starts to investigate its surroundings, we seem to sign a pledge vowing to fill our house with plastic, all in the name of baby development. Willingly or not we've all contributed to that sea of plastic. But is there an alternative? Is there such as a thing as a plastic-free toddler?

PLASTIC-FREE FEEDING

This might be the hardest nut to crack for a parent who wants to eliminate plastic from the kitchen. Despite the fact that the World Health Organization suggests it's good to breastfeed for up to two years,[21] most mums in developed countries give up around six months in when junior moves on to solids.[22] Mass-produced baby food goes as far back as the 1860s when the first formula milk appeared on the market, and which some doctors claimed was better than that of a wet nurse. Swiss merchant Henri Nestlé invented the first artificial infant food, called Nestlé's Food, and just twenty years later there were several brands of baby food, mostly made up of grains that needed mixing with milk or water. By the 1920s the first commercially canned baby food arrived and in the 1930s came the first precooked dried baby food, originally made for sick children. We've come a long way since then, although some might not think the flavour has improved . . .

Many women can't wait to get their firstborn on to solids. Until a few decades ago, when mums were moving their babies on to solid foods, they could get teeny tiny glass jars containing anything from chicken and rice, chicken and vegetables, rice pudding with apple, banana custard, cheesy pasta and roast dinner. Looking forward to this next step, many counted down the days until baby was ready to wean. Babies mostly dived into their exciting new meals and our bins rattled with tiny glass jars.

Fast-forward thirty years and all that has changed. Although glass jars do still exist, when it comes to toddler food packaging, single-use plastic is king!

When you've got a toddler, and certainly if you've got more than one, convenience is your best friend and so single-serving plastic pouches are fabulously handy. Plus, they encourage your toddler to feed themselves. A squeeze of cottage pie followed by a suck of yoghurt – lunch sorted! Glass jars need a spoon and they're not as convenient to cart around if space in the nappy bag or toddler back pack is at a premium. But glass jars are a great option if you're going plastic free as they can be recycled. If you have time you could make your own baby food and use the jars for storage in the fridge or freezer – just remember that since the lids' seals are broken they won't be completely airtight and leave space at the top so the glass doesn't crack as the food freezes.

Here's another thing you might have noticed: cutlery. Toddlers seem to come with an endless supply of plastic knives, forks and spoons. Does your toddler only eat with a spoon that's adorned with Fireman Sam or My Little Pony? And when that phase is over, they move on to using a fork covered in Paw Patrol or Elsa. Why not simply go straight to stainless steel cutlery – it's cheaper and easier and plastic free. And it's just a spoon, right? Who needs a fancy handle? Have you ever bought sets and sets of plates and cutlery for your toddler but last time you looked only the knives were left? Forks and spoons seem to constantly disappear, so then you have to buy another set!

Why not start your toddler off with proper cups and plates also? You'll need to be more vigilant, but you can get heavy-bottomed ones that are pretty hard for your weaning six month old to throw across the room.

Feeding a plastic-free baby at home is, you will discover, a relatively simple job. You'll just need to plan a bit more. It can get a bit trickier when you've got two toddlers and there's hardly time to breathe. When parents are in overload mode, friends or grandparents can help by making batches of baby food and popping them in the freezer in reusable containers (if you're using plastic containers, at least they won't be single-use). Invest in one of those tiny choppers that blend everything, from cottage pie to the Sunday roast, into creamy meals.

It's actually when you leave the house that trying to have a plastic-free toddler becomes a bit of a nightmare. It's probably down to the fact that because you have so much to do just to get out of the door, food is the last thing on your mind. And shops and restaurants know this so they've filled a gap in the market. Now you can buy plastic-packaged baby/toddler food almost anywhere – from yoghurts, squeezy pouches and drinks in a plastic bottle to the extra-tiny plastic pots containing a piece of flapjack or some carrot sticks. It's almost all single-use and hard to recycle.

Here are a few tips to avoid some of that nasty single-use plastic on a day out.

- Take a reusable bottle full of fresh water . . . easy to refill pretty much anywhere.

- Pack some fruit, like bananas or chopped-up strawberries, in reusable containers for a quick snack.

- Bring your own home-made food from your fridge or freezer with you and if you need to sit down somewhere to eat, ask for it to be heated up. It is quite rare to find a place where they would refuse to do this for you.

So ridding your baby/toddler menu of plastic takes a bit more thought but it's very achievable and, once you're in the groove, pretty straightforward.

TIME TO PLAY!

Plastic toys invading the house is a very common sight. Plastic is durable, hardwearing and colourful and it can be bendy or rigid. It's the gift that keeps on giving when it comes to playthings.

We're all about telling the world how bad plastic is for the marine environment because it never fully goes away, just gets smaller and smaller, entering the food chain right at the bottom and then making its way up. So with that in mind . . . is it ever safe for your baby or toddler to suck on a bit of plastic? Apart from the choking hazard associated with the tiny bits, are there any further consequences?

Your toddler's push-and-go plastic car of today could be yesterday's old car dashboard or redundant printer. Sounds excellent, all that recycling, right? But it's the plastic recycling process that could potentially be harmful when it comes to toys. For instance, Hexabromocyclododecane, or HBCD, is a brominated flame retardant substance that was commonly used in the 1970s. It's so toxic that it can no longer be sent to landfill or recycled and in 2013 over 150 nations agreed to phase it out after studies suggested it affected thyroid function and brain development.

However, a study by IPEN (a global civil society network) and Arnika (an environmental organization in the Czech Republic), found that recycling plastics that contain the likes of HBCD resulted in the contamination of some children's toys, including the Rubik's cube, which is supposed to be good for brain development.[23]

Modern toys marketed in the European Union are approved under the Toy Safety Directive. The Directive states that certain chemicals cannot be used in the accessible parts of toys and that a number of heavy elements like mercury and cadmium are also not allowed in parts of toys that children can get to.[24] It's great news if you're buying new, but many parents get hand-me-downs and buy second-hand, because it makes financial sense. But there is no regulation covering the recycling or resale of older toys.

In early 2018, scientists from the University of Plymouth analysed 200 used plastic toys found in homes, nurseries and charity shops across south-west England. They found that old building bricks, dolls and toy cars contained materials that didn't meet modern toy safety guidelines.[25]

Barium, lead, bromine, cadmium, chromium and selenium were discovered in many of the toys they investigated – all of which were of a size that could be chewed by young children.

So although passing things on and buying second-hand can often be a good thing, you may need to give it some serious thought when it comes to toys.

The good news is that it's not all about plastic when it comes to toys. One of the alternatives is wood, ideally unfinished wood which has just been beeswaxed or organically oiled. Wooden puzzles are great (but avoid walking barefoot unless you want to experience excruciating pain) and wooden rocking horses remain an all-time favourite. You can even get unfinished wood bathtime toys, which you just dry out on the side once the little ones are in bed. You can buy pull-along wooden toys, wooden play garages, wooden and steel musical toys, wooden kitchens and workstations with proper wooden tools. (Although it has to be said that many have tried the wooden tool set but in comparison to its plastic equivalent, judging from the expressions on their little faces, it's bo-r-ing.) Now, there's no denying it, these toys are going to be slightly more pricey. As with all toys – you do need to do your homework when buying, especially off the internet. Quality, safe, wooden toys, manufactured by ethical and environmentally conscious companies are available and most will highlight their standards on their websites or packaging. But they will last and last and can be safely handed down and are in some cases – like a wooden play kitchen – a thing of beauty. The marks and dents will tell a tale of someone's childhood.

The thing is, when your baby starts on their toy journey, everything heads to their mouth. And it's much easier to wipe away dribble from plastic and give it a quick rinse under the tap. There is good news from the world of toy-making though. Some manufacturers are exploring the use of what they describe as more sustainable plastics compared to petroleum-based products. Some combine natural fibres or wood flour with bio-based plastics to create durable water-resistant toys.

Letting a child's imagination run riot can lead to a world of early years creativity. From their first handprint picture, often done at nursery, to grabbing a big chunky crayon and having a good scribble – kids love to create.

But beware . . . one of the staples of childhood creativity is an ocean nightmare: glitter! It might seem like a harmless bit of fun – it was a mainstay ingredient of the 1970s and 80s *Blue Peter* generation's 'making' kit – but it's made of plastic and there are, rightly, calls for it to be banned. It's a micro, microplastic. One group of nurseries in the south of England decided to ban 2,500 children from using it in 2017. What a great move . . . but it's still widely available in most craft shops. Use rice and lentils instead and even pasta when doing the next creative session with your kids and encourage your nursery or pre-school to ban glitter, too.

And in the grown-up world – avoid make-up with glitter in. Cosmetics chain Lush has already replaced glitter in its products with synthetic, biodegradable alternatives.

LET'S GO OUT FOR A WALK!

So you're ready to go for a walk with your enormous double buggy. It's snapped into shape by flipping a red plastic catch here and pulling a green plastic lever there, and, with the help of four other people, your easy-to-erect toddler transport system has a baby on one side and a toddler on the other. And they're fully encased in plastic. Surely, when it comes to travel, there's no alternative to plastic, is there? And is there anything wrong with a plastic pushchair anyway?

Well, yes there is, because many will have certain levels of toxicity which you'll probably want to avoid. At least make sure your pram or buggy isn't full of toxic flame retardants, PVC, BPA, Teflon, phthalates, and Scotchguard. Strollers, buggies and pushchairs must comply with certain British Standards, but keep your wits about you when buying second-hand.[26]

Today you can easily find non-toxic buggies, strollers and prams online at sites like Gentle Nursery or peaceloveorganicmom.com.

READY FOR A BATH

After being out and about, after the snacks and the games there is no better way to end the day than with a nice bath. Unfortunately bathtime is a whole new plastic minefield, just like it is for grown-ups. Shampoo, bubble bath, lotions for dry skin, pots of cream for the nappy area. You name it, it's in plastic. But look hard enough and you can swap to zero-plastic alternatives. Baby Bee Buttermilk soap bars, for instance, are totally wholesome and there's no plastic in sight. Dove also does a baby bar if you want to go more mainstream. Lafe's do baby shampoo, lotion and body wash packaged in what they describe as baby-safe plastic.

5 THINGS YOU CAN DO TODAY TO DETOX CHILDRENS' LIVES FROM PLASTIC

1. Use washable, reusable nappies – not disposable ones.

2. Pack your own drinks and snacks when you're out for the day so you don't have to buy plastic-wrapped food. Use recyclable and reusable glass jars.

3. Ditch the cartoon character cutlery and move straight to stainless steel.

4. Swap plastic toys for health-proof wooden toys.

5. Have a plastic-free bathtime with baby soap bars and unfinished wooden toys you can dry on the side.

NOTES

....

....

....

....

....

....

....

....

....

....

....

....

....

....

'With just ten minutes of preparation the night before, you can leave behind the processed cereals and plastic-wrapped breakfast bars.'

4

BREAKFAST

08:00

For many of us, mornings are a bit of a rush, and are
often filled with unforeseen events – missed alarm,
spilled cereal, lost school shoes and the rest. It can be
difficult to make time for a proper breakfast. We've grown
up with the standard fare of cereal and sliced bread, even
though their nutritional value is often questionable. It's
good to make time for a tasty and healthy start to the day
and with a bit of planning, even a quick morning munch
can be a pleasure, rather than a chore. Plus, by making sure
you eat something before you rush out the door, you'll be
able to avoid all that plastic that can come with grabbing
something on the go. Here's how to have a plastic-free start
to your day.

THE FULL BREAKFAST

Whether scrambled, boiled, poached or fried, eggs are cracked open every morning in homes across the land. Although some come in a plastic or polystyrene (non-recyclable) box, eggs in cardboard boxes and recycled paper pulp trays are easily found (these can even be put on your compost pile as they are biodegradable). Chickens make delightful pets, too . . .

If you eat meat, all varieties of bacon, sausage, black pudding and liver, and other delicacies, can be bought without plastic wrapper at the butchers and the supermarket fresh meat counter. Go prepared with a tin or reusable plastic Tupperware or other suitable container to take it home in.

Mushrooms, tomatoes and spinach can be bought fresh and loose, and beans can be bought in recyclable tins. Avocadoes come in their own compostable packaging!

Fried bread can help make sure you don't throw away any slightly old bakery products (although it won't help your arteries).

MILK

We've got used to the high-density polyethylene (HDPE) bottles that contain up to several litres of fresh milk at unfathomably cheap prices in the supermarket. The bottles, while not made from reused plastic, are at least recyclable. Fortunately, the possibility of a home delivery in glass bottles hasn't completely died out; in fact, home deliveries are enjoying a revival. If you live in a large town or city, you can probably have a daily delivery of fresh milk in glass bottles at a very reasonable cost. Check out findmeamilkman.net, or other websites online. If not, there are lots of uses for plastic milk bottles in the garden (plant containers and watering cans, for example) or the garage (for containers for hazardous oils to take to the recycling centre) – look online for a hundred and one ways to reuse a plastic bottle!

FRUIT JUICE

Like milk, fruit juice now comes in HDPE bottles or Tetra Pak containers. Doubtless it is just as good for you as anything sold in a glass bottle. But why not do your own juicing? A freshly squeezed fruit juice cannot be beaten for flavour and nutrient content; if you don't have the time to do this, it can be delivered to your doorstep along with your milk.

FRESH FRUIT

Always buy fruit loose – this is easily done for robust fruit like apples and pears, bananas and citrus fruits. Soft fruit doesn't travel well without being damaged in transit, and is routinely packaged on the farm in a hard plastic container and wrapped in poly film. As well as strawberries and raspberries, cherries, blueberries and redcurrants (which are more robust) often come triple-wrapped, somewhat unnecessarily. One way of avoiding all this packaging, although it won't work for out-of-season and imported fruits, is by heading to a nearby farm to 'pick your own'. It's a fun and effective way of buying fruit, even getting a little exercise at the same time. Old-fashioned fruit and veg shops are scarce nowadays, although some do exist. Markets pop up in all towns and cities, particularly at the weekend, and give you the chance to buy fresh fruit in season, which you can take home in a paper bag or, better still, loose in your reusable bag.

TEA AND TOAST

Hands up, who knew that teabags were made of plastic? Until recently teabags were always made of paper, and although the occasional tea-leaf explosion would occur if you handled the bag too roughly, they seemed to work fine. Now many teabags contain polypropylene, and we aren't given any warning. If these bags are thrown on to the compost pile, plastic will be spread around the garden. Fortunately, many brands are going back to paper again – so look out for them. Loose tea is also a great option.

For toast, one plastic-free option is to make your own bread. If that's a bit of a stretch, you can pick up your loaf direct from the local bakery without a bag or plastic wrapping. Sliced bread is a bit harder to contain, but for uncut loaves, that stay soft and fresh for longer than sliced ones – a paper bag will do. If you do buy a plastic-wrapped loaf, think about reusing the wrapping for sandwiches.

You can slather jam or marmalade on your toast as long as it's from a glass jar with a metal lid. It can be a bit of a challenge to find plastic-free butter or margarine. Stay away from anything in plastic tubs or wrapped in plastic-lined foil. You can get butter wrapped in paper-backed foil or paper but it can be hard to tell. Ask at the counter or supermarket helpdesk.

CROISSANTS AND COFFEE

Coffee beans can be found in non-plastic pouches or bags for home grinding, particularly from specialist coffee shops. Ground coffee for filters and percolators is usually sold in foil-coated plastic bags, too. However, there are lots of bespoke coffee makers who offer plastic-free coffee varieties online. Check out Roasting House (roasting.house) but there are other options, too. Perhaps surprisingly, some of the instant coffee brands come in all-glass containers with glass lids too. If you have a coffee machine that uses capsules, check with your brand to see how they recycle. Coffee capsules or pods are not usually accepted in your home recycling collection but some brands have their own system.

A French breakfast – that is, breakfast in France (and several other European countries) – is usually made with fresh ingredients: from the freshly baked baguette, croissant or brioche and the freshly ground coffee to the freshly squeezed orange juice. Cheeses, cured meats and fruit are also often part of the meal. So, go continental – make the effort to have a fresh start to each day. Just make sure you buy all your ingredients plastic free.

CEREALS

Breakfast cereals shout their health benefits from their cardboard cover but inside the box you'll often find a small plastic bag filled with the wheat, oat or rice grains that are baked with added vitamins thrown in – along with added sugar and salt. Mueslis and granolas can be some of the most sugar-loaded breakfast options. Any parent of young children will know cereal somehow gets into every corner of the house each morning, too.

There are some salt- and sugar-free branded cereals available, but they're not plastic free. You can buy unbranded cereals, such as cornflakes, from some places, such as health-food shops, as well as new 'plastic free shops' which are starting to open up throughout the UK. You simply scoop what you need into your own container. Some mueslis and granolas come in all-cardboard packaging – just sometimes with an unnecessary plastic window to look at the product through. With lots of other options for breakfast, you could easily give cereal a miss. Or at the very least buy in bulk, which can really help to cut down on the amount of packaging.

PORRIDGE

Oats are one of the few morning staples widely available without any plastic wrapping as they usually come in a simple cardboard container. Whether you like it sweet or salty, made with water or milk (from a glass bottle, of course), porridge makes a top plastic-free brekkie. Plus, oats are cheap as chips.

You can make a jar of your own plastic-free Ready to Heat-and-Eat Porridge; it beats the high fat and sugar quick-pots from the shops every time. Simply soak the required amount of porridge oats in milk in a jar, with lots of lovely berries or some nuts, sweeteners like honey or syrup, and leave them overnight. In the morning, you can eat it cold, or heat in a microwave (perhaps in the workplace kitchen if you're running really late) – and you're ready to go 'til lunchtime!

Home-made muesli

500g oats

100g chopped dates

300g mixed seeds

100g chopped mixed nuts

100g raisins

2 teaspoons cinnamon

Mix all the ingredients together and store in an airtight container.

BREAKFAST ON THE GO

Too often there's just no time to make a really good breakfast, and to eat it. Like with all other meals and snacks, you can pick up ready-to-eat cereal bars or muffins, but most will have a wrapper made partly or wholly from plastic. Your takeaway coffee might seem like it comes in a recyclable cup but most coffee cups currently aren't recyclable. If you do need to have a coffee on the go, there are lots of portable, reusable options now available (for example see uk.keepcup.com or andkeep. com). Choose one and simply ask your coffee shop to fill it up. One supermarket has even announced that it will only serve their free coffee to customers who bring a reusable cup.

For munching on the run, picking up a piece of fruit is always a good option or perhaps something freshly made – a bagel, pastry or croissant from the bakery. A home-made oaty breakfast bar is both delicious and plastic free. You can buy porridge cartons if you're really caught short – at least the cardboard packaging is degradable.

5 THINGS YOU CAN DO TODAY TO DETOX YOUR BREAKFAST TIME FROM PLASTIC

1. Try to find time to cook breakfast at home. Or, if you know you'll be in a hurry, be prepared – make your breakfast the night before to take with you in the morning to avoid grabbing something while you're out.

2. Get to know your milkman – book your milk and fruit juice delivery at just the time you need it.

3. Check out your local market for fresh seasonal produce that you can buy loose. Or even pick your own fruit and veg at a nearby farm.

4. Invest in a portable coffee cup and ask your coffee shop to fill it up instead of using a disposable cup.

5. Know your Ts and Cs – tea and coffee all have a plastic price tag when it comes to convenience, so look out for varieties that come without the plastic attached.

NOTES

....

....

....

....

....

....

....

....

....

....

....

....

'Three pounds, that's all it takes to kill the sea. A sandwich boxed in plastic, a pack of crisps and a plastic soda: the (bad) meal deal.'

5

THE OFFICE

09:00

For most of us in the UK, the workplace is in an office. Here, you might be tempted to think there's not much you can do: so much of what you see is either in plastic or plasticized: chairs, desks, computers, printers, carpet, ceiling. Some get lucky and work in a place with wooden floors and wooden tables (and windows that allow the circulation of fresh air!) but they're a small minority.

However, you mustn't lose hope because wherever you work, there is always something you can do. Does your office have a green policy, with actions to combat plastic? NO? Head straight to the office of the CEO and make it happen. There are lots of eco options for the stationery you use. Need a ring binder for a report? Cardboard ones are available, and look sleek. Desk tidy or in tray? Cardboard again. Pencils, pens, rulers, mouse pads – all come in a range of materials, including wood, cork and recycled plastic.

But there is one particular moment in the day when your choice can make a real difference. It is that very precious hour: your lunch break.

KILLING THE SEA FOR UNDER £3

Three pounds: that's all it takes to kill the sea. Forget for a moment everything you've read about beauty products, pollution, climate change and so on. If you want to find the true source of plastic excess you must venture to the shiny temple of our times: the supermarket.

The quest for a plastic-free office life needs to start, in particular, at the on-the-go lunch aisle at the supermarket. It features a 'meal deal': sandwich, crisps and drink for under £3. Just £3 . . . how do you beat that? Food has never been this cheap and this convenient.

On-the-go lunch is a corollary of contemporary urban life. And although the naked ape's first natural habitat was the vast African savannahs, urban life seem to be 'the future' for the human species. According to a report compiled in 2014 by the Population Division at the United Nations the big day happened at the beginning of this decade: for the first time on Planet Earth, the majority of human beings lived in a city. It's a seemingly unstoppable trend: in 1990 there were just 10 megacities (those with 10 million inhabitants or more); in 2014 there were 28, hosting more than 450 million people. And by 2030 there will be 40. In case you're wondering, the ranking is led by Tokyo with 38 million inhabitants, and central London hardly makes it into the league with 9.7 million.[27]

....

We are so urbanized and used to urbanization that the news went largely unnoticed. And wrongly so, because urbanization is a disaster for nature. Those millions of individuals, crammed into concrete buildings, need an incredible amount of energy, food and water and produce a mind-blowing amount of litter and sewage. Cities are the very opposite of the whole zero-miles philosophy because nearly everything that is consumed by its inhabitants is produced somewhere else.

Cities, however, have always been a synonym of progress. Putting lots of minds (ideas) and capital together in one place, has produced amazing results, both immaterial (cultural movements) and material (technology, science). Not to mention social innovation. For a long time, the city – that unnatural, messy and busy concentration of human souls – was, and in some areas of the world still is today, the only place where those from LGBTQ* and BAME† backgrounds, as well as women, can escape prejudice, discrimination and male domination. And then there are immigrants who bring with them an incredible mix of cultures, colours, sounds, flavours, religions and traditions, producing cultural innovation in cities around the world.

Cities are to the human species what reefs are to marine life: a temple of diversity. And on Planet Earth diversity and life are synonyms: you can't get one without the other.

* Lesbian, Gay, Bisexual, Transgender, Queer
† Black, Asian and Minority Ethnic

So cities are not all bad, which is why today a large number of organizations are working to make them more 'sustainable'. One obvious place to start? The 'meal deal' aisle. Lunch back at home is mostly a thing of the past, and this 'meal deal' is as devilish as it gets.

JUST A SANDWICH

The sandwich is in a plastic container, of course, because customers like to see what they're buying. Some packagers have tried to reduce the amount of plastic and they've come up with a cardboard box with just a side of see-through plastic. The result is that the box is unrecyclable, one of those cases where 'the road to hell is paved with good intentions'.

There's a whiff of good news here – one leading supermarket has taken the simple step of making the plastic film easily removable from the cardboard – via a peelable tab – meaning there's no reason why the cardboard can't be recycled.

The contents of the box are just as bad, and probably worse. A sandwich can only be that cheap if it uses very cheap ingredients. The salad, tomato, bacon and chicken are probably the product of the main destroyer of life on Earth: contemporary agriculture, in particular agriculture that relies on the intensive use of natural resources and lots of chemicals (fertilizers and pesticides). There is a whole encyclopedia on the side effects of our broken food and farming industry. And even if you're not an animal welfare-ist you cannot escape the horrors that are caused by some aspects of contemporary industrial agriculture. Pollution from pesticides, fertilizers and animal-farm runoff kill life on land and in the sea just as much as, and probably more so, than plastic. The intensive use of land requires growing quantities of organic or inorganic fertilizers. If these nutrients reach the sea they can trigger massive blooming of algae that absorb all the oxygen, creating 'dead zones' for sea life. According to a recent study these de-oxygenated 'dead zones' have quadrupled in volume in the last fifty years.[28]

That cheap sandwich is the final product of a system that is trashing forests, ploughing up grasslands and river deltas, and driving species to extinction. And it's not just happening in the Amazon; it's happening right here, in the UK. The plastic wrapping is nothing but the cherry on top of an environmentally toxic cake.

A PACKET OF CRISPS

Then there are crisps: how can you not love crisps?

Unfortunately, the plastic packets they are packed in are incredibly resilient and will float in the sea – poisoning and choking sea life – for centuries. They have featured in our Great British Beach Clean top ten ever since we started it twenty-five years ago. Recently a journalist's attention was captured when one young beach cleaner picked up a packet of crisps that was older than he was (they could date it because the brand had not been in business since the early 90s). One careless snack on the beach . . . decades ago.

Home-made crisps

1. Boil thinly sliced potatoes for 3–5 minutes.

2. Drain well and combine with oil (sunflower or olive oil)

3. Scatter with salt, pepper and either garlic powder or dried oregano or basil.

4. Place on a greased baking sheet in one layer.

5. Bake at 230°C for around 15 minutes until golden brown and crispy.

A DRINK IN A BOTTLE

And then there's your single-use plastic bottle. Last year a report on global packaging trends by Euromonitor International revealed that we consume a million plastic bottles every minute. In 2016 a whopping 480 billion bottles were sold, and numbers are rising.[29] If you think about it, it's shocking but not an inconceivable number: it's an average of 50 bottles per year per person. But the reality is that if you're a Londoner, or other big-city dweller, and particularly if you're a 'meal deal' fan, you probably buy 200–250 plastic bottles per year.

And buying your lunchtime thirst quencher in a small carton may initially seem like a good option. They are recyclable. But the little plastic straws attached to them are not and straws are among some of the most frequently found litter items on UK beaches. To date, nobody has come up with an alternative straw that will fit in the carton packaging, be bendy enough to drink from and strong enough to break through the little foil seal – and that's also recyclable.

HEALTHY LUNCH

Of course, many 'on the go' lunchers have decided to choose a healthier option by cutting out bread and choosing salad. Supermarkets, coffee shops and even garages have every salad combination you can think of on offer . . . and it's all encased in a plastic box and served with a little plastic fork. Healthier it may be for you than a double egg and bacon bloomer, but the packaging is yet another choking hazard for the ocean.

The top two items found on the beach during last year's Great British Beach Clean were bits of plastic and polystyrene (225 items found per 100 metres) and food packets, such as crisps, sweets, lollies, sandwiches, etc. (40 items found on average per 100 metres). To put this in context, this is more than one item per every step on the beach. Caps and lids came in at number 5 (one every 5 steps) and at number 10 plastic cutlery, trays and straws.

This 'meal' might look like a 'deal' to you, but it is a true environmental disaster. All for under £3.

5 THINGS YOU CAN DO TODAY TO DETOX YOUR OFFICE FROM PLASTIC

1. Choose stationery made from cardboard, wood, cork or recycled plastic.

2. Find out what your office, other workplace or school is doing to combat the use of plastic and for the recycling of it.

3. Avoid plastic-wrapped sandwiches and other food. There are plenty of delis that will make you a fresh baguette or panini and wrap it in paper. Or just carry your Tupperware with you.

4. Go refillable: use flasks or reusable plastic bottles. You can find fancy ones that will filter your water. Water fountains are increasingly common. You'll save quite a lot of money, too!

5. Most importantly: take your time. Plan what you want to eat, buy (or grow!) the ingredients plastic free, and prepare your food at home to avoid the dreaded meal deal and all that unnecessary waste.

NOTES

....

....

....

....

....

....

....

....

....

....

....

....

....

'Disposable cups and glasses are one of the biggest pollutants in the world because so many are used.'

6

AFTER-WORK ACTIVITIES

17:00

Your back and shoulders hurt, and your figure is not quite
what it used to be. That's what a life in an office can do to you.
And that may be one of the reasons why, according to the
2017 State of the UK Fitness Industry Report, some 10 million
of us (1 in 7) are a member of a gym.[30]

Most of us are neither athletes nor gymnasts and fall in the
category of 'trier'. We know that physical exercise is both
enjoyable and good for us. We like joining in all kinds of
sports. The irony is that the activities we do to get ourselves
fit and keep healthy involve a lot of material that's less than
healthy for the planet. This is an area where leaving plastic
behind can be truly challenging.

Try to think of one sport where there isn't a plastic-based
material involved. It isn't easy. Even the winter sport of curling

(or 'Olympic housework' – just watch the frenetic brushing!) uses synthetic fibres for its brooms, Teflon for the slippery shoe soles, and shiny plastic handles for the stones.

Some of us aren't gym-goers, preferring the great outdoors where there is more to see and do, and only head indoors to the gym when winter training is just too miserable. Once inside, the gym blur of colourful activewear can be eye-watering. Lycra (also known as Spandex and elastane) is very versatile, comes in many colours, and is used in all sorts of sports clothing.

It is hard to believe that Lycra was only created in the 1950s, through experiments by the DuPont company, and in particular a hard-working chemist called Joseph Shivers. After almost having his work programme to find a synthetic alternative to rubber shelved, he discovered a plastic copolymer (made from two types of plastic material) with which he could make a thread to create a wonder fabric with many possible uses. The fibres can be manufactured in seemingly any colour, and woven with a consistency that means even the thinnest (careful – it can stretch to almost transparent!) sample can be very strong and long-lasting, while giving that comfortable, stretchy, breathable feel. Everything from horse riding jodhpurs to dance hot pants can be fashioned with this very modern textile.

Lycra sports clothing is incredibly close fitting, can be very forgiving and incredibly comfortable. There's no doubt that it is really well-suited for sport. Stretching is an essential component of working out, and Lycra stretches with you.

Most athletes wouldn't do without it now. It's easy to restrain from buying new versions each season as it lasts forever – it is much stronger and longer-lasting than rubber, although cheaper sportswear can become a shapeless and baggy embarrassment, and isn't a great investment!

Apart from the option of going naked, which apparently the original Olympians did, there are some alternatives to Lycra, polyester, nylon and the rest:

- Cotton and wool can be both lightweight and versatile, if perhaps not as long-lasting. Wool in particular may be a little more cumbersome. But Roger Bannister ran his sub-four-minute mile in a cotton vest, and many athletes train in garments made from merino wool.

- Hemp, soy fibre and bamboo exercise clothing is available, too. It all needs care to keep the natural fibres intact and looking good. But that is the point – it shouldn't go on forever. It won't be swallowed by the fishes in fifty years' time. While Lycra isn't single-use, it certainly isn't degradable.

It's great that lots of team kits these days – for hockey, rugby, netball, football and more – are often made from recycled materials. All gym wear, from headbands to socks, and everything in between – including the essential sports bra – can be made from a material crafted out of recycled PET plastic (the type that plastic bottles are made from). This is very positive, although fibres will still become a hazard when shed or washed off in the washing machine.

Training shoes are a must for a lot of sports. They haven't been around forever; the sporty ancient Roman, for example, would have worn boot-like sandals made from a single piece of leather, held together with hobnails, which protected the toes from chariot wheels and the like. These were not cheap, and most competitive sports (often to the death) would still have been undertaken barefoot, as it was for ancient Greek runners. Leather was the main material used for footwear until modern times; the leather football boots used in the days of England's victorious 1966 World Cup campaign must have been a burden before anyone even kicked a ball.

Today's trainers and boots probably feel like slippers in comparison. They are super lightweight and comfortable, but they are often made from layers of glued and stitched synthetics. Some do have natural rubber soles, like the first 19th-century 'sneakers', while others use rubber material from recycled car tyres, and some have material uppers made from recycled bottles. Look up eco and ethical sneakers online – there are plenty of different kinds. Most aren't set up for specific sports, and it's harder to find eco versions of trainers that take the hard impact of running on roads and rocks. Whatever shoes you choose, look after them. Use them until they wear out. If you're an elite runner, and need to change your trainers every few hundred miles to keep a spring in your step, donate the old ones to a shoe bank.

Back at the gym, you'll be struck by the fact that clear plastic cups are provided next to water coolers. Every person has a filled water bottle, easily replenished. But the plastic-lined bin receives countless one-use cups, all on a one-way course to landfill. Refillable bottles are the only way forward. But there are different varieties of these, too. Most cheap plastic bottles degrade, especially in sunlight, and there is evidence that materials like pthalates and bisphenol A (or BPA) can leach into the water. Steel and glass bottles are probably best, and these stay clean and look good for ages.

Exercise mats can get very sweaty and they're used by a lot of people in a busy gym, so you might want to take your own with you. Most mats are made from PVC, (Polyvinyl Chloride, for which toxic chemicals are used in its manufacture, and which is hard to dispose of safely) but you can buy ones made of natural rubber. They cost and weigh a little more than their PVC relatives but they're easy to find online. They can have a bit of a rubbery smell to start with, too, but that disappears quickly. Mats woven from jute and grasses aren't so smooth, but are fine for yoga and Pilates.

TENNIS

Tennis is a traditional game, and while more and more tennis balls are covered in nylon (a plastic), better-quality balls are coated in wool. They're readily available – every ball used at Wimbledon is a wool-dressed one. Tennis racquets used to be made of wood and the strings were originally made from catgut – actually processed intestines from sheep or cows – but these days nylon is the usual material, or other plastic wire.

CYCLING

Cycling is an increasingly popular pastime and pedal power instead of petrol power makes for a very eco-friendly way of getting around. It's also a great way to exercise without harming the joints, and recent studies show it helps retain muscle definition and a healthy immune system well into old age.[31]

However, many of us wear cycling tights made from Lycra, and a lot wear very visible padding around the bottom for a life in the saddle. Some people consider a bike ride to be an extreme adventure, needing special measures like energy gels and sweets. The advice on the wrappers is to consume the contents regularly, even every twenty minutes during exercise – the manufacturers want to sell you a boxful, without a thought for waste disposal.

Sweet and other food wrappers are consistently one of the top three items found in beach-litter surveys. Having swallowed the contents of a sports sachet, you are left wrestling with a sticky, empty pouch or wrapper whilst wobbling dangerously at speed. Too many cyclists just drop them on the floor. There are better alternatives for refuelling. Home-made flapjacks or fruitcake taste better and last longer through a ride. Or pack some fruit in your pockets (bananas have their own easy-peel compostable packaging).

Healthy home-made flapjacks

250g chopped dates

150ml water

2 tablespoons peanut butter

1 teaspoon cinnamon

2 tablespoons mixed seeds

200g oats

1. Preheat the oven to 180°C.

2. Place the chopped dates and water in a saucepan and simmer over a high heat until the dates start to break down.

3. Add the peanut butter and cinnamon and mix well. Add the seeds and mix well again. Add the oats in small increments, stirring constantly. You may need to add a little more water.

4. Grease a baking tin with a little butter and add the flapjack mixture. Bake for 15–20 minutes, until golden brown on top. Allow to cool fully before cutting into squares.

SWIMMING

Swimming doesn't need a lot of gear, but swimsuits, bikinis and trunks are usually made from synthetic plastic fibres. It is easy to find swimming shorts and costumes made of cotton or other fabrics. Towels, too, are often made from polyester, but are available in cotton and linen, which work equally well.

Outdoor swimming and almost all water sports in chilly seas and pools, invariably need a layer of waterproof fabric of some kind. In the 18th-century, the earliest attempts at diving suits were made of a cloth that was heavily oiled, but which barely kept the water out at all. Few underwater pioneers survived. When rubber was introduced to the mix in the 1850s, underwater exploring suddenly became a possibility; with the addition of a heavy brass helmet and a pipe pumping air to the brave volunteer, they had a chance of surviving being immersed in water, as long as they could breathe.

Neoprene, a plasticized rubber invented in the 1930s by DuPont, the same company behind Lycra, and other plastic polymers are now used in wetsuits, drysuits and sailing gear. It's very hard to find plastic-free wetsuits for surfing – some brands make a neoprene-free option, made from natural rubber, but they still contain a certain amount of plasticized rubber – at least 30% content. Rubber is, of course, a natural material but beware, it must be responsibly sourced or you could be using a natural material that is sourced while destroying nature.

SURF AND SKI

If you surf, wooden boards, assuming they're treated with non-toxic varnish, are available to replace the ubiquitous foam boards, which are made from a thick sheet of polyurethane or expanded polystyrene, encased in a hard layer of polyester resin to keep them from breaking. These are harmful pollutants, both for the board maker in the workshop, and for the sea.

Skis, sleds and snowboards are also available in wood, so winter sports can be plastic free.

OUTDOOR PURSUITS

Waxed jackets and oiled leather are available for outdoor pursuits like shooting and fishing. Sailors can get traditional oilskins that are completely waterproof, rather than the Gore-Tex versions, although they are more lightweight.

Whatever sport or exercise you do – keep doing it! But try to feel great and look good with as little of the synthetic plastic as you can, every time.

5 WAYS TO DETOX YOUR WORKOUT FROM PLASTIC

1. Cut down on buying new – it's always tempting to buy the newest, shiniest, most up-to-date gym kit and sports gear, but new-season designs are rarely that much better than the older ones.

2. Try non-plastic sports gear, made from wool, hemp, soy or bamboo fibres – they're more expensive, but a good investment. Try Thought clothing (wearethought.com).

3. Training shoes are hard to find without plastic – try ones made from recycled plastic, and use them until they wear out.

4. Make your own gym snacks. Its easy, plus they're usually tastier and more nutritious than any plastic-coated brand.

5. In the shower afterwards, avoid the plastic-bottled shampoos and gels, and those containing plastic Microbeads.

THE BIG EVENT

Maybe you're more likely to watch sport than play sport, preferring the excitement of a live match to a hard-earned sweat. Or maybe it's an unmissable cultural opportunity. Whatever the reason, tonight, after work, you're off to a big event.

From conferences to trade fairs, festivals to musical and sporting events, in the UK we certainly know how to put on a 'do'.

In 1851 The Great Exhibition was held in a purpose-built Crystal Palace in Hyde Park to show off manufactured goods to the public and boost international trade relations. The first Ideal Home exhibition took place in 1908 at London's Olympia exhibition centre. For a shilling, the public were educated and entertained by displays of labour-saving appliances and show homes. Reading Festival began as the National Jazz and Blues Festival in the 1960s and the first Glastonbury Festival started life as the 1970 Pilton Pop, Blues & Folk Festival.

Not only do we put on big events with aplomb, but we love to attend them, too – in our millions. In 2016 the UK events industry was said to be worth in excess of £42 billion to the UK economy.[32] That's a lot of money – and a lot of plastic. Because any event, particularly a massive one, can come at a cost to our oceans. Plastic glasses, plastic bottles, polystyrene food containers, plastic bags – all are used in their millions.

Convenience is king and plastic is the king of convenience.

But there are ways we can make our big event experience a plastic-free one. And the good news is the UK events industry is starting to make changes that will, in time, benefit our oceans.

FANCY A DRINK AT THE GAME?

The biggest problems for stadiums are drinks glasses. Disposable cups and glasses are one of the biggest pollutants in the world because so many are used. But there are companies that supply cups to sporting venues and festivals that are recycled at the end of their life into useful objects such as ice scrapers and coat hangers. Customers pay a £1 cup deposit on their first drink and then get refills in the same cup after that. At the end of the game or concert they can return the cup and get the £1 back or take the cup home as a souvenir – the cups are designed with logos and messages.

At Twickenham – the home of English rugby – on an international game day 140,000 pints of beer can be pulled. They are now being poured into the reusable Fan Cup.[33] Cricket, football and other rugby grounds have also started serving drinks in reusable cups. The Oval Cricket Ground has pledged to be plastic free within three years by introducing reusable cups and installing water fountains. If you want to make sure your next big event is at a stadium that doesn't hand out plastic, then check out their website.

FESTIVAL FACTS

The Cheltenham Music Festival is thought to be the oldest UK music event. It was first staged in 1945, followed in 1947 by the Llangollen Eisteddfod and the Edinburgh International Festival. If you're of 'an age' you may have been to Woodstock, or at least seen the movie documenting what happened in 1969 when 500,000 people descended on a field outside a town in upstate New York. At the end of the film the camera pans over the field to reveal a sea of litter stretching for miles. Back then, like now, most of the litter was drinks bottles and cans.

About a thousand festivals are organized in the UK annually, celebrating everything from literature to cooking, beer to drum and bass. Now they're not all on the Woodstock scale, but they do leave behind their own litter footprint. They are for everyone. From huge music and entertainment festivals, like Glastonbury, Leeds and Reading, to boutique literature festivals for the connoisseur, they could easily be a plastic nightmare. But like the big stadiums, many are really changing the way they do things when it comes to plastics. Latitude festival has been using reusable cups since 2009 and last year ran a bottle deposit point which collected plastic bottles to go straight to recycling.

Glastonbury says it will implement a site-wide ban on plastic bottles from 2019. It has already introduced stainless steel bottles and water refill points so any container can be filled for free. Organizers estimate that one million bottles are used at the event.

But for festivals, is it enough to just offer recyclable glasses? In the aftermath of any music festival one thing is clear . . . you may come with your tent and sleeping bag, but you don't necessarily go home with them. Modern tents are cheap and easy to erect. Thousands upon thousands are left behind every year – 20,000 in one year alone at Glastonbury, apparently.

These pop-up tents have a fair amount of plastic in them and if abandoned and not rehomed, they'll be off to landfill.

There are eco-tents made from weatherproof cardboard, which can be left behind after the event and sent off for recycling. But it really is better not to buy things for one-off use if you can look after and reuse one you can keep.

Likewise, the world's first compostable tent is also in the offing and could be available for summer festival-goers right now. It uses bio-based materials such as corn starch mixed with bamboo or silk. We don't know if this will be a good thing to use, given the amount of energy and carbon footprint that's likely to be involved.

LETTING GO

During many of life's big events, there appears to be a need to make an 'in the moment' statement.

Festivals, weddings, memorials – what better way to feel connected than to do something with lots of other like-minded people – an action. And the current action of choice for many people is . . . letting go a whole bunch of balloons or Chinese lanterns. A sight they might be, but they're also an environmental disaster. What goes up must come down and these things come down anywhere, including the beach and the sea where they pose a choking and entanglement danger to marine wildlife. Chinese lanterns are also a menace to crops and farm animals, to owners of thatched cottages, the coastguard – who can mistake them for flares, and even a problem for airports.

Chinese lanterns or sky lanterns, are small hot air balloon type things made of a paper-covered wire, rope or bamboo frame suspended over an open flame heat source. The heat lifts the lantern into the air where it can then float for miles. Once the flame goes out, the lantern starts to fall back to earth. Lantern (and balloon) releases aren't just a moment of fun. Thankfully over 50 councils are now banning mass releases of one or both on their land after pressure from the MCS 'Don't Let Go' campaign.

Between 2015 and 2016, our Great British Beach Clean found the amount of latex or foil balloons and strings on UK beaches rose by 53%.

There's an awful lot of confusion over balloons in particular, especially what they're made of and how they break down. Some people believe that because latex is natural, balloons made of it are harmless once let go. This just isn't the case. Latex can last for up to four years in a marine environment.

3 THINGS YOU CAN DO TODAY TO DETOX YOUR BIG EVENT FROM PLASTIC

1. Take your own water bottle with you and use the water refill points.

2. Ask via social media if the festival or sporting venue you're attending has a reusable-plastic plan.

3. Consider an eco-tent so if you have to leave it behind, you can do it with a clearer conscience.

NOTES

....

....

....

....

....

....

....

....

....

....

....

....

....

'There are some 51 million pets in the UK. Dogs are the most popular (9 million). They produce an estimated 1,000 tonnes of waste each day.'[34]

7

PETS

19:00

Back at home for many of us it is time to take care of our
beloved pets. Brits love them. In 2015, figures suggested that
we spent a record-breaking £7.16 billion on our animals – a
growth of 25% since 2010.[35]

We lavish our cats with plastic-coated jingle balls and we
throw plastic toys for our pooches. We house hamsters in
plastic palaces and let mice roll around the floor in plastic
balls. We pick up dog poo with plastic bags and transport our
cats to the vet in a plastic box. We love our pets and we
sure can spoil them – and much of that spoiling is done
with plastic! Is it possible to be a good pet parent without
plastic?

SCOOP THAT POOP

There's no denying it, the very worst thing about having a pet, especially a dog, is having to deal with the 'back end'. How dog owners must have cheered when the plastic doggy-poop bag was invented.

In the beginning we just left it. In 1970 the situation in New York got so bad that dog owners were told to 'Curb Your Dog' and get their pooch to do its business in the gutter, if they could find a space between the many parked cars. A scooping law was finally introduced in 1977.

In the UK dog owners are discouraged from leaving it behind with the threat of a fine. Most local councils issue on-the-spot fines to dog owners who are caught failing to pick up. However, these don't actually do the trick – a survey for Animal Friends Insurance in 2017 found that one in four Brits refuse to pick up their pet's waste, with Gloucester being revealed as the UK's worst place for leaving it behind.[36]

During the MCS Great British Beach Clean in 2011, we found piles of dog poo, wrapped in plastic bags, that could be threatening the health and safety of beach visitors. Poop scooped up into bags and left on UK beaches rose over 11% between 2010 and 2011, with Scotland recording the biggest increase – a whopping 71% in just one year.

Walking your dog in the countryside or near the coast with a pocketful of plastic dog-poop bags sounds very responsible – but it's not, if you can't be bothered to properly

dispose of the full bags and end up hanging them on a
tree, slinging them into a verge or leaving them behind a rock.

Animal waste is one of the many seemingly small sources of
pollution that can add up to big problems for our oceans and
even for our own health.

In 2017 new fines were introduced in the UK. If dog owners
are found not to be carrying at least two poop bags when out
on their walk they face an on-the-spot fine of £80.[37] But has
anybody really got the time to enforce that?

Also in 2017 MCS launched a campaign – 'C'mon, confess! Who
hangs dog mess?' – to ask people to not leave their full poo bags
behind, hanging off trees, but to take them home and dispose of
them responsibly.

But what we're talking about here is a plastic bag, and
whatever way you look at it, disposal means landfill, and that's
not a good thing.

Dog-poop bags are cheap to buy, convenient and mean
you can pick up without getting your hands anywhere
near the poop. It's common courtesy to pick up your
pooch's mess so others don't go home with it on the sole of
their shoe.

If you're a conscientious poop scooper and you know that
plastic bags are a bad thing, then you're probably buying
biodegradable poop bags. But it's not clear if these bags break
down over time as the packaging suggests. Plus, most

biodegradable bags aren't compostable – which is a common assumption.

So why not try some other alternatives which don't involve plastic bags?

- You can buy a dog poo wormery! It works in the same way as a kitchen waste wormery to create compost for your garden (but you can't use it on your veg garden, of course).

- If you've got a garden, you can bury all your dog poo in a big hole in the ground. It needs to be away from drains or water courses and about one metre deep. You can add products like septic tank activators to get the process underway.

- What about a dog loo? Now, this isn't an actual loo for your dog to sit on – there's no extra dog training to be done. Basically you sink one half into the ground, scoop your dog's poop into the 'loo' and then add a compound to break down the waste. It has a lid and most have got a child lock to make it more secure. It needs to be installed properly though, otherwise you end up with a really smelly system, attracting flies and dirty looks from your neighbours.

- The Forestry Commission, in 2017, suggested dog owners adopt a 'stick and flick' method for dealing with their dogs do dos. No bag required but you'll need to watch where you're flicking.

FOR THE CAT LOVERS

Some people, especially those who are not cat lovers, find the whole idea of a cat using a litter tray – often sited in the kitchen – pretty gross. But actually it can be very hygienic. OK, so you will almost certainly need a plastic litter tray so you're able to keep it nice and clean. You can always buy second-hand . . . it'll be plastic but not new plastic. Rethink the cat litter and make your own. Shredded paper is great and works a treat according to some cat owners.

PLAYTIME

Decades ago your dog had a ball and your cat had a jingly bell and that was the extent of your pet's playthings. Now if you go to any pet shop or browse the internet, the choice of plastic pet toys is endless.

You may think you're being a good pet parent by giving your little darling lots of gifts but, just as plastic toys, particularly old toys, can be bad for children's health, so can plastic toys be bad for the health of your pet. Chews, pulls and balls can contain bisphenol A (BPA) and phthalates. BPA and phthalates have been linked to all manner of health issues from stalling the development of reproductive organs to cancer. The US and the EU have banned some phthalates in children's toys. If you value your pet, go plastic free when it comes to toys.

Natural or recycled rubber, hemp, rope, cotton and canvas . . . there are lots of plastic-free options now available for dogs and cats. Try an eco-friendly frisbee, a hemp rope ring, recycled rubber toys or even an old shoe (a plastic-free one of course!). They may be a bit more expensive but at least you'll know that you're not just doing the planet a favour but your pet, too. They'll love you even more for it!

Don't forget the tiny pets, of course. You can buy wooden homes for hamsters . . . with lots of wooden toys. Hammy doesn't have to be incarcerated in plastic and, he can even get his daily exercise in a wooden wheel! Wooden kit for hamsters is generally made from untreated natural wood so it's safe – and very stylish. It can be a bit more pricey but it will invariably last longer.

FOOD TIME

- Avoid plastic feeding bowls and opt for stainless steel or ceramic. They last longer and you can safely clean them in the dishwasher.

- Buy pet food in bulk from pet shops by taking in your own containers for them to fill up. Avoid buying ready bagged food that has plastic liners. This goes for all pet food . . . cat, gerbil, hamster, rabbit. You can buy treats in the same way at some pet shops; they do a sort of pet treat pick-and-mix. Pre-packaged dog food has been around for 150 years or so; before that dogs ate our leftovers and butchers' offcuts.

- Avoid any pet food in plastic pouches. Tins are a better option – just remember to rinse them thoroughly before putting them in the recycling.

- You can make your own pet food if you've a mind to! There are lots of recipes and suggestions online.

Home-made dog food

Makes about 8 servings

150g brown rice

1 tablespoon olive oil

1kg turkey mince

3 handfuls of chopped spinach

2 chopped carrots

1 chopped courgette

2 handfuls of peas

Cook everything up then let it cool before dividing into portions. What could be simpler?

5 THINGS YOU CAN DO TODAY TO DETOX YOUR PET FROM PLASTIC

1. Choose biodegradable/compostable poop bags.

2. Make your own natural cat litter by shredding paper.

3. Swap any plastic toys for natural alternatives.

4. Have a go at making your own dog/cat food or at least buy in bulk.

5. Buy second-hand pet carriers, cages or litter trays – so at least you're reusing!

NOTES

....

....

....

....

....

....

....

....

....

....

....

....

....

....

....

....

....

....

....

....

....

....

....

....

'UK households produce some 31 million tonnes of waste per year, one tonne per household. It's about a kilogram of rubbish per day per person.'

8

WASTE OF TIME

20:45

The day is over. In the quiet of your home, after dinner and before some much-deserved sofa-time, someone will say it: 'Take out the bins, darling'. Whilst dragging the numerous bags to the right bin (or the bins to their correct pick-up spots), even if you're not an environmentalist, you must have asked yourself at least once: 'How do we manage to produce all this rubbish?'

And rightly so, because among the many talents of the human species, this one deserves a particular mention: our capacity to produce waste. Waste is truly a human invention since there is no such thing as waste in nature. Everything, no matter how small or big, smelly or dirty, is reused by some other organism in one big 'circle of life'. The law of nature is simple: everyone's debris becomes someone else's raw material. A wrasse will happily snack on the damsel fishes' undesired guests, its ugly parasites, and dung beetles will use other animals' excrement as food or as building material for their love pads. It's a matter of taste, of course, but the guiding principle of the natural world is that nothing goes to waste. That was until we invented, just a few decades ago, materials that would take centuries to decompose and that would, in the meantime, be beneficial to few and lethal to most species. We invented useless, damaging and poisonous litter. Litter not just unprecedented in persistence but also unimaginable in quantity.

Here are some of the 'fun facts' compiled by the waste-management company CB Environmental:[38]

X UK households produce some 31 million tonnes of waste per year, one tonne per household. It's about a kilogram of rubbish per day per person.

X 31 million tonnes = the weight of 3.5 million double-deckers, the queue of which would circle Planet Earth two and a half times. And that is every year!

X On average we produce the equivalent of our body weight in waste every seven weeks.

However you look at it, it's a staggering amount of rubbish. And this is just households, remember. Commercial and industrial waste adds another 40 million tonnes.[39]

Anyone who cares about this planet (and our health) has learned the four Rs:

- Reduce
- Refuse
- Reuse
- Recycle

Sadly, it's easier said then done. Latest available data says that we recycle only 45% of our domestic waste. The EU wants us to recycle at least 50% by 2020. Germany, Austria and Belgium are already above 60%.[40]

But where is plastic in all this? If you go through the official data you will find it in the 'packaging waste' (including paper and cardboard, too) category: 2.26 million tonnes. We actually do better than the EU average and hit our targets since we recycle or recover some 45% of our plastic waste (although we export a large chunk of it, and especially did to China before the ban). However . . . 2.26 million tonnes is only around 8% of the total amount of waste we produce so 45% seems a much better figure than it actually is.[41] And this is why: a lot of plastic does not go in the 'special bin' and ends up in a landfill (or the sea).

DEALING WITH IT

Now, it's one thing knowing the sort of single-use plastic to avoid in the home but it's quite another when it comes to recycling the bits that do slip through the net.

You'd think it'd be pretty easy – after all there are 300 different collection arrangements around England alone!

But it's this very choice and the varying systems that causes confusion and leads to reduced recycling rates because, for many people, it's just too much hard work. Too many coloured bins, too many different collections on different days. Added to this the media stories about what you can and can't recycle and things can get a little bewildering.

The Public Health Act in 1875 suggested householders keep their waste in a 'movable receptacle' – we call them bins these days. The Act also stated that councils are legally obliged to empty people's bins. In the 1930s councils were told they had to empty bins every seven days. Dustbins were heavy metal containers which didn't change their shape for decades. Until, of course, along came wheelie bins and with them recycling bins, in various shapes and sizes, from the end of the 1980s.

So now we separate our recycling into bins of varying colours: green, white, brown, yellow or blue marking out the special functions of designated wheelies, boxes, communal banks or specially designed plastic sacks. It's a new age for rubbish responsibilities.

Under EU law the UK must recycle 50% of its household waste by 2020. In 2016 Wales was already at that level, whilst England and Scotland were still struggling to hit their targets.[42]

In 2015 Fife council started a trial where waste destined

for landfill was collected just once a month in a bid to force householders to recycle more. Recycling collections during the trial were increased. At the end of the trial waste to landfill dropped and recycling increased. Sounds like a winner. Well, it may have been initially but sadly the monthly collections stopped because they were described as 'too costly' with too many people contaminating their recycling bins with rubbish that should have gone to landfill.

So back to the drawing board.

Why are so many of us struggling to get our heads around recycling? Well, let's just take everyone's favourite product – plastic. Does it all just go in the recycling bin?

Nope. There are so many different types of plastics in all sorts of shapes and sizes and grades, and it's often attached to other materials that you can't recycle. Plastic doesn't come with a clear message – recycle this/don't recycle that.

There are lengthy explanations – some bits of this can be recycled, other bits of that. Some councils will recycle this, others won't recycle that. It's not black and white. Nearly all types of plastic can be recycled, but different areas of the UK have different technologies available which means that the plastic you can recycle in, say, Birmingham, may not be recyclable in Brixham.

And even if you're sending the right things off for recycling, did you know that plastic bottles and plastic food trays can't be recycled together because they can't be used together for a new usable plastic. You can see why the cry of 'Oh, just chuck it all in landfill' echoes around some back gardens, as eyes glaze over and brains go foggy.

PLASTIC BOTTLES AND THEIR TOPS

According to the latest UK Household Plastics Collection Survey from RECycling Of Used Plastics Limited (RECOUP), bottles make up 67% of household plastic packaging collections.[43] Plastic bottles are generally made from one of four materials (grades) and these can all be popped into the recycling bin:

- PET (e.g. fizzy drink bottles and squash bottles)
- HDPE (e.g. milk bottles and detergent bottles)
- PP (e.g. ketchup bottles)
- PVC (e.g. large squash bottles) – although PVC use is in decline

All very straightforward then. Not quite, because bottles invariably have tops and that's when it gets confusing. The bottom line with bottle tops is that they can be a different plastic grade and colour to the bottle they came from. Up until recently customers have been asked to remove bottle tops – which many didn't do – because they contaminated the recycling process. Now that's not so crucial as processing plants will remove the tops. But you should swill out your bottle before you pop it in the

recycling bin. Producers are also making the bottle
tops a lighter colour, too, which reduces colour contamination.

YOGHURT POTS, TUBS AND TRAYS

Pots, tubs and trays make up 33% of household plastic packaging
collections (RECOUP, 2017).[44] Yoghurt pots made from PET
can be recycled . . . but PET isn't universal in the yoghurt-pot
world. Some are made from polystyrene, which is not accepted
in the plastic recycling process, but a pot's a pot, right? So of
course we throw them all in the recycling bin. Furthermore, in
some areas where yoghurt pots are collected, the recycling
process is contaminated because pots aren't washed out. Pots
need a swill before they go in the recycling bin.

FOOD TRAYS

Now here's a thing that may surprise you. Black food trays were
introduced because it was thought they made the food seem
sexier, a bit more exclusive and attractive to the customer. That's
right – a beef steak on a black tray makes you want it more than
if it was reclining on a white tray. All very good marketing, but
terrible when it comes to recycling. Black trays cannot be
identified and picked up using the equipment available in most
recycling sites. So black plastic trays are for the landfill bin.

MARGARINE AND BUTTER TUBS

These are called rigid tubs in the industry. They're not made
from a single polymer but often a mix. Not only that, some

tubs absorb the fats and oils they contain which means they are contaminated and even a good swill under the tap won't sort that out. So although you may think you're doing the right thing by putting your rigid tub into the recycling, you may not be. Check your local council's recycling page to see if they're accepted.

PACKAGING FILM

This can be anything from film used to (unnecessarily) pack four apples (on a black tray) from the supermarket or the cling film you use at home to keep your food fresh. Very little of it can be recycled. It's low-grade stuff and can only be made into other low-grade stuff like grey or black rubbish bags; it also gets wrapped around the processing plant machinery. Some councils do accept plastic bags and film for recycling, but they're not easy to sort and so they're costly to process. Best advice? Don't use it at home or buy food stuffs wrapped in it. There's an alternative – beeswax food wraps (beeswaxwraps.co.uk) are reusable and biodegradable – and you won't spend hours trying to find the end or you could try real greaseproof paper (but not the modern plastic kind).

PLANT POTS AND TUBS

Small plants, or plugs, ready for planting are often housed in plastic pots. You get the little plant out of the pot, and pop it in the soil but what do you do with the pot that's left behind? This is a tricky one. Is a plant pot packaging or is it a thing, a product? If a plant pot contains a plant that's intended for the ground, then it's packaging. But a pot with no plant – that may look just like the pot *with* a plant, is in fact a product. Got it? Many local authorities won't accept plant pots in the recycling. However, some independent garden centres offer a pot recycling service so do check whether this is the case in your local centre.

IF IN DOUBT, ASK A CHILD

If after this brief explanation you still feel lost you'll have to surrender to the idea of using real experts: kids. Believe it or not all children care about nature, and when you explain to them that littering hurts animals and plants they learn the art of recycling and differentiating rubbish like we never will.

MCS works in schools to run awareness campaigns on littering, and more often than not we leave the classroom thinking we have learnt more than what we've taught. It's certainly the case of the 'Bincentives' competition in which kids were asked to devise a system to encourage their fellow pupils to recycle or use a bin. Hampton High won the competition with some brilliant ideas: queue jump passes and free movie lunchtime!

GET YOUR RECYCLING BANG ON AND HERE'S WHAT WILL HAPPEN:

- Disposable utensils and meat packing can be recycled into: CD cases, office accessories.

- Bottles and jars can become stuffing for pillows, carpet backing and even sweatshirts.

- Shampoo bottles can become other bottles.

- Most bottle tops turn up as ice scrapers, and packing cases.

- Film used for wrapping meat can come back as insulation for cables and drain pipes.

5 THINGS YOU CAN DO TODAY TO BE A BETTER RECYCLER

1. Wash out your plastic bottles and yoghurt pots.

2. Check what your council accepts – there's normally a list on their website.

3. Get the kids on board – make them your in-house recycling experts.

4. Don't buy anything wrapped in unrecyclable plastic film.

5. If in doubt leave it out – contamination is a big problem.

NOTES

....

....

....

....

....

....

....

....

....

....

....

....

....

'Nothing feels quite like going to the beach on a weekend. Why not join your local MCS Beach Clean and help do your bit?'

9

THE WEEKEND

No matter how much you love your work, everyone's favourite part of the week is . . . its end. No boss, no deadlines, no crowded trains and traffic jams, unfriendly colleagues and office politics: just two whole days of freedom. The weekend is time to spend doing the things you enjoy, whether that's letting your hair down and partying, or taking it easy at the end of a hard week.

For some it starts with one precious ritual: the paper. The Saturday (or Sunday) newspaper is worth a leisurely read, and comes with an incredible array of extra supplements. Unfortunately, an increasing number now come plastic-wrapped, even though the headlines these days are often about the plastic menace. It's a good reason to try a rival publication without the plastic, if your usual comes with a polywrap.

After a good breakfast and a morning read, and before the actual fun bits, a large number of Brits take care of that weekly chore: the trip to the supermarket. These shining temples of our times are, essentially, huge depositories of plastic. The fight against useless and harmful plastic can only be won, or lost, here.

FILL THE FRIDGE – THE BIG FOOD SHOP

When the UK home nations introduced a 5p single-use carrier bag charge some people thought that the world as we knew it was over. At MCS we'd started calling for action on single-use carrier bags in shops back in 2008 and we were instrumental in getting a levy introduced across the UK. Wales was first in 2011, Scotland and Northern Ireland brought in their charges in 2014 and 2013 respectively and England came dawdling along in 2015 with a slightly diluted version of what everyone else had.

Those opposed to the charges predicted riots in the streets, shops closing at an alarming rate, angry scenes at checkouts as flimsy bags were no longer handed out willy nilly, shoppers lurching between retail outlets trying to balance an entire week's shop in their arms and a country searching for 5ps down the back of sofas, because nobody needed them for anything else, ever. But those early fears were unfounded. We've survived (and have cupboards full of bags for life). We've done it, people – we've gone some way to save the oceans from plastic pollution and it only cost us a heap of small change.

Becoming converts to bags for life, canvas holdalls or jute shoulder swingers was pretty straightforward, mainly because we had to make alternative carrying arrangements once the flimsy plastic bag was no more. And they are almost completely no more, as most of the major supermarkets now only offer bags for life. But although we may be reducing the number of single-use plastic bags that are ending up on our beaches – there was a decrease of almost 40% between 2015 and 2016 according to figures from the MCS Great British Beach Clean, the lowest number in over a decade, from 11 plastic bags per 100 metres of coastline cleaned in 2015 to just under 7 in 2016 – shopping isn't off the hook when it comes to plastic waste.

Now, you may carry bags for life everywhere you go, use a reusable coffee cup at every coffee shop and demand only paper straws in your Friday-night cocktail, but can you make the weekly shop completely plastic free?

FRUIT AND VEG

Every year, MCS asks people to take on the Plastic Challenge and try and live without single-use plastic for a whole month. It's a tricky ask and the most difficult area that challengers encounter is the food shop. We're told it's vital we try and have our five a day. Fruit and veg is a must if we're to stay healthy. The trouble is, fruit and veg is often all wrapped up snugly in plastic to, we're told, keep it fresh.

Here's the thing – a plastic film-wrapped cucumber can last three times as long as one that's only clothed in its own skin. Wrapped it'll lose just 1.5% of its weight through evaporation after 14 days, compared with 3.5% in just 3 days for a naked one.[45]

OK – let's say we'll look at letting this in as an acceptable 'single-use' – but why, being fully recyclable, is the wrapper never recycled? So, while a cucumber can last longer in a plastic wrapper before going soggy in your fridge later, it ends up causing rubbish for landfill, too. Cucumbers are provided to caterers unwrapped, because they get used up within a morning; they can be made available without the plastic. It needs us, as the customers, to buy just what we need.

But how many of us manage to get through all our fruit and veg in the time allotted? What we need to do is rethink how we shop for fruit and veg. What might be convenient for us may well be inconvenient for our oceans and beaches. Shop local, shop at farmers' markets, buy loose not packaged fruit and veg in the supermarket and don't put your loose items in those very flimsy plastic bags – that's if they haven't already run out of them anyway. Large farm shops even sell loose frozen fruit and veg and should be happy for you to take along your own containers.

PASTA

Browse the internet and look for local shops where you can buy stuff like pasta in cardboard containers. They do exist or you can make your own from fresh.

DAIRY

Why not help a struggling industry and get your milk delivered daily by a milkman! And in glass bottles, too. According to Dairy UK figures, in 1980 89% of milk

consumed in the home was delivered in the early hours
by Mr Milkman. By 2015 this had dropped to just 3% and
there were only about 5,000 milkmen and women doing a
round.[46]

Whilst we're on the subject of dairy, let's look at cheese. In the
supermarket it's vacuum-packed in plastic film or grated into a
plastic bag (buy a stainless steel grater, they're reusable!). You
can buy cheese in specialist cheese shops or from the
supermarket deli counter and avoid the plastic middle man.
Take along your own container or some real greaseproof paper
and get your favourite Cheddar wrapped plastic free. If you
buy too much you can freeze cheese. There's plenty of advice
on the internet on freezing it to get the best results.

MEAT

What about meat? In the supermarket it's usually vacuum-
packed in cling film and often sits in a plastic tray. Why not
head to the local independent butcher armed with your own
container? If you're of the Tupperware generation you're
bound to have containers of various sizes and shapes lurking
in the back of a kitchen cupboard because you always knew
they would come in handy one day. Well, now their time has
come. As long as they're clean and don't leak you can head
home with your joint knowing there's no plastic to be
disposed of at the other end.

And if you're thinking that supermarkets will never be able to
improve their plastic footprint, then think again! What's been
described as the world's first plastic-free supermarket aisle was

created in early 2018. Six hundred and eighty products, including meat, rice, sauces, dairy, chocolate, cereals, fruit and vegetables, are on offer in Amsterdam's Jan Pieter Heijestraat store of the Ekoplaza organic supermarket group. The aisle was the brainchild of British-based campaign group Plastic Planet. Everyone else take note. For now, though, you can shop smarter in the big stores, give the independents your support and check the internet for zero-waste/packaging-free shops near you.

5 THINGS YOU CAN DO TODAY TO DETOX YOUR WEEKEND SHOP FROM PLASTIC

1. Shop in local farmers' markets, farm shops and independents.

2. Buy loose fruit and veg in the supermarket.

3. Go old school and get your milk delivered in glass bottles.

4. Take you own containers to cheese shops and butchers.

5. Swap from tea bags to loose tea.

FILL THAT WARDROBE – SHOPPING FOR CLOTHES

For some it's a chore, but for others it's an addiction that needs feeding. We all need to buy clothes and it's a great excuse to head into town, to the out-of-town retail outlet or just spend hours surfing websites. During the weekend, high streets in most cities across the country are packed with people of all ages and walks of life. And if many now realize how problematic food shopping can be when trying to go plastic free, most of us don't have a clue about the amount of plastic that's almost everywhere in our clothes.

THE RIGHT OUTFIT

We've come a long way when it comes to our gear. The caveman had their animal-skin loincloths, the Greeks looked good in tunics and the Romans had their togas. Blended fabrics and mass production and the concept of 'fashion' took us to a whole new level – and then, of course, we wanted cheap and lots of it so our wardrobes became an ever-changing kaleidoscope of kit.

Clothes keep us warm (or cool), protect us from the elements and injury, identify us as being part of a group or show how important we are. Jeans, for example, are hard-wearing and long-lasting, originally invented for miners and factory workers, and cowboys to wear on long Wild West days. But they are more than just durable fabric; denim has long been, and still is, a cultural symbol of alliance, like 'teddy boy' or 'skinhead'. Today, jeans are a fashion item, worn with style, whilst still being at the top of the list for people looking for a heavy-duty fabric.

Whether you wear clothes for practicality or fashion, plastic synthetic materials have taken over in the last fifty years or so. And polyester, polyamides (like nylon) and other plastic threads, found in so many fabrics, shred into microfibres from the time they are made, while they are worn, each time they are washed and spin-dried, and then when they are disposed of. It's hard to get away from having plastic in your clothes, but here's how to go almost utterly plastic free.

UNDERWEAR

Let's start with the smalls. And briefs might seem an easy place to begin, with cotton and bamboo versions very widely available for men and women. But look at the label on even the skimpiest thong, and you may be surprised to find it's made up of a mix of as much as 40% polyester. It seems that materials used by almost all underwear manufacturers are based on a polyester weave; even those of pure organic cotton still use elastic made from petroleum-based rubber, which gives the close but comfortable fit. Bras are engineered with significantly more synthetic stitching and yarns.

What can you do?

Well, you could make your own briefs .... but would you want to? There are some small online and mail-order companies who promise organic cotton-, bamboo- or hemp-only smalls.

HOSIERY

The invention of nylon in the 1930s brought stockings and, later, tights to the wider world, after real silk stockings had been the preserve of the wealthy. Women went wild for them, and some 64 million pairs were sold in 1940 alone; demand outstripped supply, and black-market prices rocketed.[47]

While the glamour of having smooth legs encased in a variety of colours is less of a novelty today, it's still a formalwear and workwear staple. Over 200 million pairs of tights and stockings are sold each year in the UK alone – none of which can be recycled.[48] Even wool tights for colder weather are usually a wool blend with, you guessed it, nylon or polyester. But pure merino wool varieties are around – they just cost an arm and a leg – well, both legs really. Going bare legged is by far the easiest option, although usually one for warmer weather.

FORMALWEAR

Suits and dresses made in traditional ways and from natural materials are easy to find because they are made for special occasions, and events where people need to look their best. It's actually a pleasure to look through well-made, luxury garments that fit well and make you look and feel fantastic. A full lounge or dress suit in wool, with cotton shirt and silk tie, is available off the peg for men, without the need for a tailor. Sumptuous dresses and work suits come in linen, cotton, wool, mohair, tweed and many others – be adventurous, but as always, check the label and be ready to spend just that bit more.

LEISUREWEAR

Formalwear is easy to get hold of plastic free because it's that traditional clothing style that has been around for centuries, using materials from a pre-plastic era. Leisurewear is an altogether newer concept – and of course we want it cheap. Many of us want to look like the latest celebrity sportsperson or popstar, but we have various physical attributes, so leisurewear needs to be versatile.

The 1980s saw a boom in leisure fashion, with high-street outlets and leading brands alike jumping in with sweatbands and socks in bold pinks, yellows and greens, multi-coloured sweatshirts, and, the ultimate leisure fashion item of yesteryear, the shell suit.

The use of modern fabrics in our everyday clothes persists. It's not just their favourite local football club's kit that every fan has to buy. Or the Lycra sports leggings that make the wearer feel fitter, faster and stronger whilst changing TV channels. Jerseys, sweaters, jeans, tee-shirts, breathable outdoor clothing – it's all manufactured, on a mass scale, in synthetic or mixed natural-synthetic fabrics.

DISPOSABLE FASHION

The cost of clothing has become so minimal that many people buy new garments to use just once or twice, and then throw them away. A little black dress, accessories like handbags and hairbands – they are often so cheap compared with the cost of a dinner, or even the taxi ride home, that they're thought of as disposable. It's a worrying trend that is encouraged by pile-them-high-and-sell-them-cheap department stores. The quality of the items is so poor that they don't look good for long anyway. It is the absolute opposite of the care previous generations took over their clothes and other belongings. And it's a shocking waste of resources. Changing the mindset from disposable fashion, that's cheap to produce and buy, to robust fashion that lasts, is desperately needed.

OUTLET SHOPPING

Outlet shopping became popular in the 1990s and is still doing well. Last season's designs are sold for a sizeable reduction in price, still drawing in customers for the brand name, and the bargain. They can be a plastic-fee winner because they offer the plastic-free shopper the chance to seek out some of the eco-conscious garments that seemed prohibitively expensive when they first hit the rails.

VINTAGE FASHION

New fashions are leaving behind the values of the past. Retro fashion is the antidote to the disposable trend. Whether bought new or second-hand, the vintage look is a statement to say that throwaway just isn't on. Just be careful to give the retro-look plastic imposters a miss.

JEANS

Buying denim clothes, and especially jeans? Don't buy stretch denim – it contains a small amount of Spandex woven in to help it stretch, but it will lose its shape more quickly than a pair bought to fit well in the first place. Even though sagging jeans have been in fashion in the past, and maybe will be again, they are never a flattering look!

TEE-SHIRTS

Like jeans, the humble tee-shirt started with plain beginnings. Initially designed to wear under something rather than on top, they have become a steadfast in most wardrobes. Traditionally made out of plain cotton, these days they are embellished with designs through the use of Plastisol. This is a PVC emulsion mix, invented around 1960 for use as a coating in industrial products, and which takes and holds colour incredibly well. The dyes are used on hooded tops and sweatshirts, as well as tee-shirts.

Plastisol will come off on a steam iron and mark all of your clothes if you're not careful. If you want to wear tee-shirts with a design motif, think more along the lines of those printed with water-based inks. Or try other printed fabrics, perhaps ones that are dyed with the ancient Japanese technique of Shibori from around 600 BC that uses dyes made from plant leaves and roots to make lovely screen prints.

FLEECES AND HOODIES

Many of us have them, as they're lightweight, warm, versatile and easy to wear. They have replaced the woollen jacket and overcoat in one fell swoop. Most contain virgin polyester, but sometimes they're made from recycled PET plastic from bottles. Most zips used on fleeces, hoodies and coats are manufactured by a single company, and appear all to be based on a synthetic polyester fabric, too. Go back to wool, like cashmere or merino – they are both warm, they last well and are lightweight. Cotton hoodies without zips for the cooler months, are also a good choice.

FUR OR NOT TO FUR?

It's come back full circle – fake or faux fur is popular again. First introduced in 1929 its fakeness was ideal for people who didn't want the real thing. Made from modacrylic and acrylic polymers, the fibres fall out and float, in air and water.

FOOTWEAR

The ethics of footwear are the same as for clothes, leaving you to do what you can where you can, and where your conscience leads you. Vegans have been at the forefront of alternative footwear, with retro espadrilles and plimsoles, and more besides. But many shoes sold as vegan are made of plastic, whether recycled or new.

Woollen uppers have been used in many shoes, including trainers, but it's the comfort soles that contain plastic – usually moulded ethylene vinyl acetate (EVA). Leather soles are uncomfortable, and natural rubber versions are hard to find and less hardwearing.

Buy for the longer term and to make sure your footwear fits its purpose; high-quality, well-made footwear will last. A really good pair of trainers or hiking boots may last you years and cut down on waste.

Recycle the kids' shoes, as these are worn for shorter periods of time. Children are more likely to grow out of their shoes before they wear them out. Pass them on!

WASHING

All clothes need washing and drying, and it's here that tiny pieces of material, individually too small to see, get shed by clothes as they tumble in machines. This makes a very visible fluff you'll see in a tumble drier filter; look closely, best with a hand lens or microscope, and you'll see tiny strings and ragged fibres that were once part of clothing fabrics.

If you have put together an entirely plastic-free wardrobe, you can be sure that the fluff will biodegrade. For most people, though, the tiny microfibres will be plastics, their long-lasting, lightweight properties meaning they'll drain into water treatment works, and straight through to rivers, streams and the sea. Estimates vary as to how many or how much plastic gets into the environment this way, but a single clothes wash can release as many as 700,000 microfibres – a staggering amount.[49]

So far, no washing machine or dryer manufacturer has developed a filter to stop these fibres draining away, and fibres fall off your clothes as you wear them daily anyway. It makes it all the more important that we cut out the plastic where we can.

5 THINGS YOU CAN DO TODAY TO DETOX YOUR CLOTHING FROM PLASTIC

1. Inspect the label: when buying items, check that it is as close to 100% natural material as possible – even if it says 'wool rich' and 'made from 100% organic cotton' this can often be just a proportion of what's in the fabric.

2. Don't buy cheap: whatever the material, cheap fashion doesn't last. It will lose shape rapidly, fall apart, and become waste material in a short time.

3. Go retro: old-style garments made traditionally look great and last for ages – but avoid clothes made to look retro that are actually plastic imitations.

4. And second-hand: it's better for the environment than buying anything new, and the fact it still looks good means it should be a long-lasting item.

5. Don't do disposable: party to go to and can't think of what to wear? Wear something you already have in your wardrobe! You do not need to buy something new for every special occasion.

THE WEEKEND TAKEAWAY

Who wants to cook on the weekend? A takeaway meal, perhaps delivered to your door to eat in front of the telly, can be a real treat, and it's definitely the easy option. But it often comes contained in various plastics. Pizza usually comes in a flat cardboard box, so can be a reasonably safe plastic-free bet – but if the cardboard is contaminated with food, remember that it can't be recycled. Fish and chips, too, are often paper-wrapped, depending on where you buy them. Chinese and Indian takeout tend to be the biggest plastic fiends, separating each item in a lidded plastic container. Some do offer Thali (platter) service in reusable pots. You could encourage your local takeaway to use foil trays, or even agree to fill your own reuseable containers. Eating out is usually less of a plastic-filled experience – just watch out for the needless plastic straws in drinks.

THE GARDEN CENTRE

The weekend is a great time to engage in a passion. Gardening is a popular weekend pursuit that falls somewhere between work and pleasure. Serious gardeners can avoid the plastic pots, trays and compost sacks by producing their own plants in home-made compost. This takes space, time and effort, which few people have to spare. So, most of us head to the garden centre to pick up plants and all of our gardening paraphernalia. Here, plastic abounds. Most plants come in a plastic pot. You can sometimes find plants sold in terracotta pots, and you can certainly buy these on their own. They are a little more expensive, but will last and look good for years.

If you buy the plastic, return any pots you won't use again to the garden centre or nursery where you bought them; they should be able to reuse them.

Sheets of weed-control fabric are sold in rolls, tempting the buyer with an easy option without explaining that they are made of plastic fibres, which are bound to enter your soil when they rip and tear. You can use all sorts of biodegradable material as a mulch to suppress weeds, including cardboard, woodchip or leaves. Don't bother with plastic plant labels: used wooden lollysticks, or dried twigs are just as good and can be marked in pen or pencil. Some seed catalogues sell wooden labels and seedboxes, if you're happy to shop mail order.

GOING AWAY

The weekend is often the time to spend with family and friends, sometimes in faraway places. Whether you get there by train, bus or car, plastic packets, cups and wraps are in every buffet carriage, kiosk and service station. You need to put some planning into the journey, taking plastic-free provisions, including water bottles, with you. Buying a gift on the way? A bunch of flowers is always welcome, but can you get one without plastic wrap when you're in a hurry? A local florist will usually have loose flowers ready to arrange, and perhaps provide paper to wrap it with – or you can take your own newspaper for the purpose.

Getting away from it all often means a trip to the country. It's amazing how much modern farming uses plastic, and how obvious it is in the countryside. Bales of silage are wrapped in polythene, and secured with twine made from polypropylene. Those strawberries you pick up in the supermarket, in a plastic tray, in a plastic wrapper? They are grown under tunnels of sheet plastic to control growing conditions and help the fruit ripen early in the season. If you take a walk to the top of a hill with views over open country, you'll often see a shimmering plastic layer on several surrounding fields. Some soft fruit is planted in growbags of black plastic, grown for one or two seasons, then replaced with a fresh stock of plants in fresh plastic bags. Farmland can seem wrapped in plastic. In China, 12% of its farmland is mulched with the stuff.[50] Over here, a trip to the countryside can be an eye-opener to the sheer volume of plastic in use, and might influence your food-buying habits!

THERE'S NOTHING LIKE THE BEACH

Few things will create that sense of freedom we yearn for on the weekend like a day by the sea. Scientists and psychologists are still trying to understand why the simple sight of the ocean calms and soothes and puts us in a better mood. One thing you can do on a day out at the seaside, one that we certainly recommend, is join a beach clean. And chances are those people with litter pickers and tabards already on the beach are staff or volunteers with the Marine Conservation Society.

For many of us, a day in the office involves a pretty uncomfortable chair, hunched shoulders over a keyboard, a hovering mouse, a whirring printer, ringing phone and a steady supply of coffee to get you through to 5pm. It offers the same view every day in a temperature held at a steady 18 degrees regardless of the weather outside. And, of course, it's crammed full of plastic. For some staff at the Marine Conservation Society, their office is a far cry from the above. Every day there's a different view, sometimes it's hot, sometimes cold, often wet, but it can be sunny and dry, and boring office carpet tiles are replaced with sand or shingle. Their office is the beach. It's where they work. But it does have something in common with your conventional office – it, too, is crammed full of plastic.

What a miserable thought that is. You see we're using the outdoors as one big dustbin. The latest data from the annual Great British Beach Clean revealed a 10% rise in UK beach litter between 2016 and 2017, with litter from eating and drinking accounting for up to 20% of all rubbish that volunteer beach cleaners found.

The beach is no longer – well, perhaps it never was – a pristine stretch of sand or shingle. Was there ever a time when the only thing between your toes after a seaside stroll was sand? Now a gentle walk will result in bits of plastic, cotton buds, bottle tops and polystyrene sticking to your soles. You only need to do one beach clean to realize that the beach is a plastic paradise. It's everywhere you turn. So don't get all jealous of MCS staff who call the beach their office, because right now our coast is a clean-up job that seems never-ending.

A beach clean may sound like a bit of fun. Warm sun and sand, a litter picker and job done. If only it were that simple. For MCS staff it's literally a full-time job and that's before they even set foot on the beach. First of all, you have to get people to sign up to a clean. You can't do the mass cleanings we do with just one man and his dog. You need to engage with people and plenty of them. It's a never-ending round of social media posts, emails and promotion by our volunteers on the ground. MCS runs cleans all year round. Some are led by our staff and others by our fantastic band of volunteers. The Great British Beach Clean is our flagship event, held every third weekend in September. It's the UK's largest and most influential national clean-up and survey of Britain's beaches. All the items our volunteers find are registered according to an internationally agreed methodology.

In 2017 we got a massive boost from the excellent *Sky News Ocean Rescue* and the fabulous BBC *Blue Planet II* series. Sir David Attenborough's narration brought the issue of the plastic pollution of our oceans into our front rooms. People who up until then may not have given it a second thought, or even a first one, wanted to do something to stop this environmental disaster.

People care. They want to help. Maybe they did before but they didn't know just how urgent things were. Now they do and, thankfully, they're helping at our beach cleans in their droves. It's great to see 'first timers' rock up ready to clean. And the great thing is that there isn't a 'type'. We see people from all walks of life, young and old, male and female. And that is probably because right there, on the beach, we can all make a difference. Gardening gloves, litter pickers, bin bags and a survey form later and MCS staff give them a snapshot of the issue that's lurking beneath the waves, on the strandline, in the rock pools and on the beach. They're surprised to say the least. You know how people behave when something bad has happened? They whisper together in hushed tones. Well, it's not dissimilar down on the beach. People are shocked. They tell you they're shocked, surprised, appalled. They also say – right before they tell you how shocked they are – that 'it all looked so clean'.

In 2017, the Great British Beach Clean saw just under 7,000 volunteer beach cleaners pick up record amounts of litter over four days from 339 UK beaches – a staggering 718 bits of rubbish from every 100 metres they cleaned. Tiny bits of plastic and polystyrene were top of our list of finds, but there are others that we also have a real opportunity to fix.

Wet wipes, dog-poop bags, plastic bottles and plastic bags have all been reduced or made the focus of public attention due to successful campaigns for MCS and other organizations. But there's a whole swathe of items that need to be tackled.

Everybody hates litter. 'I love all that litter on the side of the road and the beach,' said nobody, ever. And nobody admits to leaving litter behind them or throwing it out of the car window: 'Remember when we went to Cornwall and left the entire contents of our picnic on St Ives beach? Oh, how we laughed!' It just doesn't happen . . . well it does, obviously, but we don't acknowledge it. We hold up our hands to smoking, drinking, even a flutter on the side . . . but littering? Never.

So where does the plastic cutlery, foil wrappers, straws, sandwich packets, lolly sticks, plastic bottles, drinks cans, glass bottles, plastic cups, lids and stirrers that we find on our beach cleans blooming well come from? We found 138 pieces of this type of litter, on average, per 100 metres of all the beaches cleaned and surveyed by our volunteers in 2017.

All the remnants of human life seem to arrive on a beach at some point. You don't believe it? Let us take you through life on the beach – you'll be amazed!

Christmas

Christmas on the beach like they do down under? We could just about do it with what we've found: an artificial tree, fairy lights, festive sticky tape, decorations and even a TV to watch the Queen's speech!

Here boy!

Dogs love a beach walk, but it seems their owners want to leave behind more than paw prints in the sand. We've found leads, collars, bowls, ball throwers, loads of tennis balls and, of course, bags and bags of dried-up almost-fossilized faeces.

Beauty and the beach

You would not believe the number of beauty items we find on beach cleans. Just about everything you need for your beauty regime somehow ends up at the coast: toothbrush, razor, dental floss, comb, hair bands, hair clips, perfume bottles, lipstick, mascara, contact-lens holders and, of course, heaps and heaps of wet wipes.

On the raz

Do most nights out end up at the beach? Apparently, seeing as we've collected: socks, pants, bras, earrings, high heels, belts, beer bottles and lots and lots of drinking straws. How does this happen?

5 THINGS YOU CAN DO TODAY TO DETOX YOUR WEEKEND FROM PLASTIC

1. Avoid newspapers with plastic-wrapped supplements.

2. Choose takeaway options that come in cardboard or foil, or collect it in your own reusable containers.

3. Grow seedlings from scratch and use biodegradable materials to suppress your weeds. Take your plastic plant pots back to the garden centre for them to reuse.

4. Plan journeys so you don't have to buy plastic-packed snacks from garages, trains and service stations, etc.

5. Join in with a beach clean! Go to mcsuk.org/beachwatch/greatbritishbeachclean for more info.

NOTES

....

....

....

....

....

....

....

....

....

....

....

....

'Buffet cars, airport lounges and motorway service stations all have at least one thing in common: plastic convenience packaging.'

10

HOLIDAYS

Whether you want your vacation time to involve extreme
activities, partying or simply relaxing and unwinding,
you can spend a fabulous time away while giving plastic
a break, too.

The ethical traveller is faced with lots of dilemmas.
Firstly, where to go: how green is the place you're staying
in? What impact does your choice of lodging have on the
local environment? And, of course, how to get there:
flying is certainly not eco-friendly but road and rail have
their own environmental impacts, too. We need to fly less,
certainly, but there are limited options if you need to travel
abroad. So adding in trying to do something about the
global plastic problem can seem pretty daunting. Happily,
while holidaying plastic free can involve a bit more prep,
there's a lot you can do, easily and simply, to ditch the
plastic.

TRAVEL ESSENTIALS

Whatever your travel plans, there are some general rules that you can apply:

- Pack light: there's almost never any benefit to taking more clothes, gadgets or home comforts than you really need. How many times have you come back from holiday having not touched half the contents of your suitcase? If you can fit your essentials into a carry-on bag, all the better.

- Pack your own bag: a cotton drawstring bag (available in some shops or there are lots online) is almost unnoticeable, but you can use it to take a book, sunglasses and suncream out to the pool, or to carry a few mementos back from a local market. Cloth tote bags are also useful and come in a range of sizes.

- Travel essentials: a travel kit needs just a few hygiene items, like paper-wrapped soap, bamboo toothbrush and a shampoo bar (see Chapter 2 for more inspiration!), and it can all fit into a small cloth bag.

- First-aid kit: take a minimized set for emergencies. It doesn't need a green plastic zipper case. While items that might be used on a wound, or that are for the eyes, need to be sterile, most things don't need to be wrapped in plastic. Some medication can be obtained in a glass, or at least recyclable plastic bottle instead of non-recyclable blister packs. Calico bandages don't need a wrapper.

- Last-minute panic buys: avoid anything sold as 'disposable', whether plastic or not – choose items you'll value and want to use again.

PACKING

It's essential, but can be stressful. Not many of us enjoy packing. Most of us have thought at one time or other: 'If only someone else could do it for me' – but you know you wouldn't want others to leave important items behind. So you pack. And you always pack too much, making sure everything you could possibly need is crammed into one case.

It is really hard to find a large, sturdy but lightweight suitcase that's not made of synthetic nylon, polyesters and the rest. Few materials can withstand being packed and zipped up tight, then loaded into plane holds, shuttle buses and cars, and be light enough to carry, too. You can buy a heavy-duty, cotton canvas duffel bag with around 75 litres capacity, to carry on your shoulder. Large leather cases are available, but are bulky. Keep using and reusing any luggage you have – wheels can be repaired, rips and tears stitched or taped up.

Cabin luggage doesn't have to be so robust, and there's lots of choice for natural-fibre handbags, holdalls and carry-on cases. Jute, hemp and cotton make soft, light options that can be squeezed into tight cabin lockers. Rigid cases made from aluminium are available which can hold a laptop and other precious items very securely, but they are expensive; you can pack computers, handheld devices and cameras in a normal carry bag padded with soft clothing.

GETTING THERE

Buffet cars, airport lounges and motorway service stations all have at least one thing in common: plastic convenience packaging. Polystyrene coffee cups with plastic lids, bottles of water, snacks wrapped in poly film – avoiding the need for these by bringing your own plastic-free provisions is definitely the best option.

This is especially important for drinking water, which you can take everywhere except, of course, beyond airport security control. Fortunately, several airports provide a source of drinking water flight-side, so make sure you take a refillable water bottle with you. While they are often well-hidden, a quick question to airport staff will provide details of the whereabouts of a water fountain. If one doesn't exist, you can just ask at a café. It isn't being cheap – it saves on the plastic.

SUN PROTECTION

It's really crucial that skin is cared for, keeping away the sun's harmful UV rays. The saying 'prevention is better than cure' applies here – once you're burnt, it's really already too late.

Sunblock lotions and creams are widely available, but they are almost always sold in a plastic tube or bottle. You can get skin creams in glass jars with metal lids, sometimes found in the baby section of the pharmacy. Some suncreams come in a cardboard tube, and although the tube can't be recycled being coated in cream, it will decompose more quickly in landfill than plastic ever will.

UV suits for small children are stretchy, plastic-fibre based suits that keep the sun off sensitive skin. They are perhaps an acceptable investment to keep children safe, especially if you can pick up a second-hand or hand-me-down one. Toddlers outgrow clothes of every kind so quickly that they rarely get the chance to wear out. Pass your old ones on to a friend or charity shop rather than throw them away.

ON THE BEACH

It's while on the beach that many of us have the vivid realization that waste plastic has spread far and wide. Nobody wants to bathe on a dirty beach, and it is so easy to avoid adding to the problem while we're there.

With your swimwear, look for a brand which uses entirely recycled materials – there are few ideal non-plastic alternatives around. Buy natural rubber (responsibly sourced) flip-flops for footwear, or, again, there are plenty of recycled options.

Sunglasses can be picked up with plastic-free frames, made of metal or even wood. Lenses, especially polarizing ones, are usually plastic though – if you can do without them, great, or buy ones you'll love and cherish forever! Make sure whichever pair you choose has UV protection.

Inflatable lilos are prone to puncture, can easily float off in the current or when there's a breeze, and don't really warrant taking with you.

Buy food or ice cream fresh over the counter, rather than in wrappers, and you'll leave only your footprints behind on the beach.

EAT AND DRINK LOCAL

As well as keeping food miles to a minimum, buying foods that are local to wherever you're visiting will often involve less plastic than familiar branded snacks, or western-style fast-food franchises. Street food is popular in lots of places, and really worth a try – buy cooked foods made from fresh ingredients that aren't likely to carry harmful bugs.

Tap water can be a potential source of tummy upsets when you're in a new place. If you're only staying a week or two, it may be best to buy a large store of water to refill your bottles each day. There are various options available to the traveller for sterilizing water, but they are not to everybody's taste. Ask your doctor for advice.

Tourist bars and clubs can be very plastic-heavy. Go grown-up with your cocktails, refusing the straw, the mixing stick, test-tube shots and parasol or sparkler (even if not plastic, they're single-use and wasteful). Wine glasses should be just that – made of glass.

HOTEL LIVING

We've all come to expect a certain level of hospitality when staying in a hotel, resort or bed and breakfast. OK, just occasionally it can be helpful to have a toothbrush provided at reception when you've forgotten yours and all shops are shut. But the plastic tends to be overdone in places – the miniature toiletries, plastic cups, and sometimes even the toilet seat is wrapped up in it. Some hotels and chains are more responsible, and do without the plastic freebies without compromising on cleanliness. Support them!

CAMPING

Camping is a great way to enjoy nature, living in the outdoors and enjoying wild surroundings. But tents and camping paraphernalia can involve a lot of plastic, generating waste that could spoil the very environment that we love to spend time in.

Staying dry and warm, sleeping comfortably, and cooking and cleaning are the main challenges of camping. Tents need to be light, portable, waterproof but breathable, and so are almost always made from plastic. Modern carbon fibre tent poles are a composite of plastic, and the guy ropes are nylon. Only the last items to go in – tent pegs – are usually metal, taking the strain of being hammered into hard ground better than plastic.

There are some plastic-free tent options. Canvas, made from cotton, linen or hemp, is tough and long-lasting. You can buy a canvas tent from specialist equipment stores, but they can be bulky and heavy, and are not all that practical an option for a family trip. They are just right for ready-made camping and 'glamping' sites, some with tepees and yurts, where you can move straight in to a (hopefully) leak-free and sturdy home away from home without the challenge of pitching it yourself.

Cheap camping chairs are made of a long-lasting plastic material, but their frames break easily. Foldable wicker stools are a lightweight, plastic-free alternative. Most sleeping bags, sleeping mats and airbeds are comfy because of their plastic ingredients. Using them again and again is one option. Or you could try a camp bed, made with a light, foldable wooden frame, and cotton hammock cloth – although bulkier than plastic mats and beds, they are very comfortable. Sleeping bags made of cotton and wool are available, but costly and not hugely practical in very wet or humid conditions.

SELF CATERING

Self catering does involve the same chores as at home, but it's an affordable option, especially for larger families or groups. Like at home, plan ahead, buy in bulk, avoid the ready-meals – you'll have a tastier and healthier experience for it. Use the same types of cleaning materials you do at home, and bear in mind that some places have a very different recycling regime, if at all, to what you may be used to where you come from.

GOING LAST MINUTE

We often leave it to the last minute to book a weekend city break or rural escape. But you can still stick to the same plastic-free rules that you'd apply with a bit more planning – have a plastic-free travel kit at the ready for whatever choice you make, for example. This will help you to avoid buying last-minute essentials that always come in the dreaded plastic.

5 THINGS YOU CAN DO TODAY TO DETOX YOUR HOLIDAY FROM PLASTIC

1. Plan ahead and take your own water bottles and snacks with you when you're travelling so you can avoid eating on the go.

2. Look for ethical swimwear brands that are made with recycled materials.

3. Eat and drink locally – avoid western-style food outlets and eat like the locals do.

4. Avoid plastic straws, stirrers and other cocktail paraphernalia if you're out in the bars and clubs.

5. Look at 'glamping' sites as a camping option as these will often have already-erected canvas tents or yurts.

NOTES

....

....

....

....

....

....

....

....

....

....

....

....

....

....

'From crackers to chocolate eggs, most celebratory paraphernalia are encased in plastic – or they pop out a plastic gift that nobody ever needs.'

11

SPECIAL OCCASIONS

There are a couple of occasions that thrive on plastic –
Christmas and Easter. From crackers to chocolate eggs, most
celebratory paraphernalia are encased in plastic – or they pop out
a plastic gift that nobody ever needs . . . a pocket-sized plastic
bottle opener for instance. Actually, maybe that's a bad example!

So how come two ancient religious festivals are now owned
by plastic?

EASTER EXCESSES

The egg is a symbol of new life, fertility and rebirth. For
thousands of years, eggs have been decorated. The ancient
Sumerians and Egyptians of 5,000 years ago would decorate
ostrich eggs, and place them on graves to represent death and
rebirth. The Christian custom of eggs at Easter started in
Mesopotamia when eggs were stained red in memory of the
blood of Christ that was shed at the crucifixion.

You could say Easter got commercial in the 17th and 18th centuries. Egg-shaped toys were made and sold to children, often filled with sweets, and the first chocolate Easter eggs are thought to have been made in Germany and France at the start of the 19th century. The first chocolate egg in the UK was made by J. S. Fry of Bristol in 1873. Now the UK chocolate Easter egg market is worth in excess of £220 million with, on average, 80 million Easter eggs sold annually.[51] That's a lot of eggs and a lot of plastic because Easter is the new Christmas when it comes to overdoing the plastic. With some brands, it's plastic packaging gone mad.

But as with any other area of life, you can break the Easter plastic habit. It just takes a bit of planning. And there's plenty of time to plan seeing as most Easter eggs start to appear just after Boxing Day – at least that's how it feels!

HOW EXACTLY CAN YOU HAVE A PLASTIC-FREE EASTER?

Moulded plastic is everywhere at Easter. You need to keep that egg in place in its box, so what better way than to encase it in clear moulded plastic? Then there are bendy plastic windows and, of course, hundreds of thousands of little plastic bags of sweets or baby eggs either inside or very close to the main-event egg.

A *Which?* investigation found the best-selling Easter eggs have an average of 25% packaging![52]

THINGS YOU CAN DO WHEN EASTER APPROACHES TO REDUCE YOUR PLASTIC IMPACT:

- Choose eggs wrapped in foil and packaged in cardboard not in plastic.

- Buy a foil-wrapped bar instead of an egg (this may not go down too well with the kids, though!).

- Buy smaller, loose eggs as they are less likely to be over-packaged.

- Buy some silicone moulds and make your own eggs!

DIY eggs

You'll need:

Easter egg moulds which you can buy at most high-street kitchen-equipment shops or online.

Good-quality chocolate: 2–4 100g bars will do, depending on the size of your moulds.

1. First, temper the chocolate. To temper the chocolate, you will need a cooking thermometer, a heat-proof bowl and saucepan of hot water. Break the chocolate into small, even pieces and melt gently in a bowl over a saucepan of hot, not boiling, water. Place the thermometer into the chocolate and heat until it reaches 43°C. Take off the heat and cool to 35°C.

2. Pour the chocolate into the moulds until they are filled.

3. Pop into the fridge until firm.

4. Remove from the fridge and then peel the mould off.

5. Secure two egg halves together by briefly placing the edges on to a pre-heated baking tray so they melt slightly, and then stick the edges together.

6. Result: your own home-made Easter egg!

CHRISTMAS CHAOS

It's the most wonderful time of the year, but Christmas is now cocooned in plastic everywhere you look – from plastic trees to blow-up snowmen, strings of little lights to turkeys encased in cling film and reclining on a plastic tray.

HOW DID CHRISTMAS BECOME A PLASTIC FEST?

We have the Victorians to thank for today's commercial fest! They really exploited the season of goodwill to within an inch of its life. They spotted that the generosity of spirit that flowed in December could be turned into a marketing and money-making opportunity. Christmas shopping was born.

Gordon Selfridge turned the Christmas window into an art form and invented the concept of the shopping-days countdown to Christmas. Department stores created Santa's grotto and queues formed as children begged to sit on the knee of the jolly bearded one. The First and Second World Wars interrupted the commercialization of Christmas but it all got back on track in the 50s when Brits were told they'd 'never had it so good'. Come the 70s and Christmas TV ads became a thing; now we expect the commercial orgy to start just as the kids go back to school in September.

We want our Christmases crammed full of stuff. For just a few weeks we want cheap and tacky and we're happy to deck our halls, not with holly, but bits of plastic in various festive shapes and sizes.

But we can have a plastic-free Christmas and here's how:

THE DECORATIONS

Buy a real tree. It's fun to choose your tree and if you get one in a tub you can plant it out in the garden as an all-year-round festive reminder. Or buy a wooden one – they make a very striking addition to your festive lounge.

Make your own decorations – there are lots of ideas on the internet and if you've got young children they'll invariably bring something made from dried fruit home from school for you to treasure for many years!

The winter landscape is dominated with reds and greens, so get out there for some natural decorations rather than buying them. Sprigs of holly and fir, twigs and branches, mistletoe and pine cones all make beautiful additions to your Christmas home.

But even if you've got a shedload of decorations made out of plastic – just hang on to them. The concept of needing a new festive colour scheme every year is just a marketing ploy. Keep hold of your old stuff.

PRESENTS

There are lots of options here.

- Go wooden, go second-hand, rustic and handmade. Swap the Christmas high-street crush for the traditional Christmas market.

- For the make-up fan there are biodegradable eye shadows that come in a cardboard compact (leave off the glittery ones); wooden toys for the little ones; books are always welcome; craft beers in cardboard tubes. They're out there and they're plastic free!

- 'Experience' presents are great, too – give your loved ones something to look forward to.

- Christmas isn't complete without some goodies for the bathroom, so make sure your gifts are Microbead-free.

WRAPPING PRESENTS

Gift wrap contains plastic so give your Christmas presents that nostalgic feel with recycled brown parcel paper, and use biodegradable eco packing tape. Add some natural string, as Sellotape contains plastic and is certainly single-use. Finish it off with some seasonal greenery and use last year's Christmas cards for tags and you'll end up with unique, beautifully wrapped gifts. Shredded paper is a great alternative to polystyrene to protect breakables.

THE BIG FOOD SHOP

Try shunning the supermarket for the farmers' market and local independent shops. You'll be able to buy loose fruit and veg, and your turkey plastic free. In the mad rush don't forget to take your reusable bags on your trip to the shops!

Get crusty loaves plastic free from the bakers, and take your own airtight containers to the meat and fish counters.

Christmas puddings are tricky – they often come in an easy-to-microwave plastic bowl – but you could always make your own. It is a bit more time consuming but very satisfying. Mince pies are cheap as chips from the supermarket but can come encased in plastic. Again, make your own! They're so simple and so tasty – and it can even work out cheaper.

Christmas is a big cheese event. Give as a gift as well as serve after pud . . . but make sure they're wrapped in real wax paper. Or take a couple of airtight containers to your local deli and just ask them to pop the cheese straight in. Whilst you're there, look out for some chutneys in glass jars. Grapes are difficult to find plastic free, but you could try your local greengrocer.

THE OFFICE PARTY

It's that time of year where throwaway plates and plastic cups make an appearance at the Christmas do. Proper plates with real cutlery and glasses will not only make a classier get-together, but you'll dramatically reduce your plastic waste, too.

CHRISTMAS COCKTAILS

If you're celebrating with a tipple or two, avoid plastic straws and stirrers. Instead, add to the magic with mint leaves, rosemary, cranberries and orange slices. Freezing your autumn blackberries in ice-cube trays makes for a great addition to your winter cocktails.

If you're a big fan of straws, then try stainless steel or glass ones or have paper ones on hand if you're likely to be visited by little ones.

More and more bars are going plastic free so be prepared to dip the lip!

ENTERTAINING

Ready-made canapés come encased in particularly large amounts of plastic packaging. But there are so many quick and easy nibbles you can make, plastic free:

- Crudités

- Cheese twists

- Potato skins

- Sweet potato wedges

- Sun-dried tomatoes

- Stuffed olives

- Bruschetta

- Home-made crisps

Any many, many more!

Christmas dips

Avoid the standard plastic-packed party mix dips this year, and make super-quick, plastic-free alternatives.

Sour cream and chive

Mix foil-wrapped soft cheese with mayonnaise from a glass jar. You'll be amazed how much this tastes like sour cream! Simply add some chopped fresh chives.

Pizza dip

Mix foil-wrapped soft cheese with a couple of spoons of pasta sauce from a jar for a dip of cheesy, tomatoey goodness.

Home-made hummus, guacamole and salsa are all great plastic free options.

SAD TIMES

There are times when, in everybody's lives, sad and tragic events happen. At such moments, finding a way of commemorating and celebrating a life is of enormous importance, making concerns about things like plastic seem insignificant. But it is worth thinking about the consequences of how to celebrate life, and preserve memories – and it's likely that no-one would wish for environmental harm to be a legacy of a funeral or other commemorative event.

Flowers are always welcome and show you care. To be really thoughtful, bunches of blooms need to come without plastic, as plain stems or, untied, wrapped in paper. Plastic wrappers soon look unsightly, and stay that way until they blow or wash away, or are cleared up and put in the bin.

Setting off helium-filled balloons as a remembrance, especially for children, is an emotional and pretty sight. But setting off balloons can lead to unintended consequences. While some pop into tiny pieces when they reach a great height, those that stay whole will drift afar. They've caused deaths of grazing farm animals, and have been found swallowed by marine turtles and other wild animals. The plastic ribbons attached to some balloons entangle birds, even as they fly. Balloons are great indoors for decorations but – please – don't let go.

Likewise, Chinese lanterns, while looking amazing, are a real hazard, too. They've been blamed for fires in buildings, and the frames last a long time as litter on fields and beaches.

5 THINGS YOU CAN DO TO DETOX YOUR SPECIAL OCCASION FROM PLASTIC

1. Make your own chocolate eggs when Easter approaches. They're fun to do and can make lovely gifts. Get the children involved and they'll forget all about the plastic-packaged shop-bought varieties.

2. At Christmas time, buy a real tree and source your decorations from nature – holly and ivy and fir cones are particularly Christmassy.

3. Brown paper packages tied up with string: wrap your Christmas presents up old style and decorate with some seasonal greenery. They will look very striking – just make sure the present inside is plastic free, too!

4. Do your Christmas food shopping in your local independent shops and at the farmers' market. Take your own containers with you to the deli and ask them to fill them up with your cheese selection.

5. When you're entertaining, serve some home-made canapés and dips rather than buying them from the supermarket where they'll most certainly be encased in plastic. Your guests will appreciate the effort, too.

NOTES

....

....

....

....

....

....

....

....

....

....

....

....

'Gloves, scrubs, gauze and syringes: single-use plastics save lives in hospitals.'

12

PLASTIC THAT SAVES LIVES

'The man laid out on the operating tables of our hospitals has more chances of death than the English soldier on the fields of Waterloo.' This was said by Sir James Simpson, Scottish genius and pioneering surgeon in the mid 1800s. And it was no hyperbole.

A colleague of Simpson's, surgeon Joseph Lister, later pointed at some injured soldiers and recommended that they should be taken 'out of this house of death, to an empty clean house with fresh air and they'll survive'.[53] The 'house of death' he was referring to was the hospital.

The idea that a hospital is a place where people's lives are saved is a recent one. Hospitals used to be dirty, smelly chambers of death until about a hundred years ago. It took geniuses like Simpson, Florence Nightingale, Pasteur, Kock, Semmelweiss and many more to turn them into what they are today. And then plastic arrived, and we entered the age of contemporary medicine. Well, actually it all started with a simple rule: wash your hands.

VIENNA, 1847: THE HOSPITAL WHERE NEW MUMS WERE DYING

Hungarian doctor Ignac Semmelweiss knew that he had to do something about the fact that the women in the hospital who had just had a baby were dying like flies of childbed fever. In the obstetrical clinic the infection was killing three times the number of women than on the midwives' ward. Semmelweiss knew that it was to do with doctors' hygiene. His colleagues were visiting patients, performing surgeries and autopsies, and then visiting female patients in the obstetrical clinic without washing their hands with disinfectant. He tested a new practice for a while: before entering the clinic doctors had to wash their hands with a lime chlorinated solution. Mortality among the young mums dropped to 1%. He could not explain it though, since bacteria and other microorganisms that cause lethal infection in wounds had not been discovered yet.

His colleagues, however, could not accept the idea that they might have something to do with the deaths so despite the evidence they discredited doctor Semmelweiss, and he was eventually fired and locked up in an asylum, where he died soon after, insulted and treated like a charlatan. Poor Ignac was right, of course. Scientists discovered the existence of microorganisms that could, through contaminated equipment, bed linen or even airborne droplets, infect patients' wounds, leading to death.

Heroes like French microbiologist Louis Pasteur provided support to the germ theory, disproving the doctrine of spontaneous generation (of infections) that was held at the time. Pasteur's work proved that without contamination microorganisms could not develop. A hundred and fifty years later we still use his technique (called pasteurization in his honour) to kill harmful germs in food and drink, such as milk and wine.

The point is that once our protective layer of skin is damaged, because of a wound or during surgery, we are incredibly fragile beings. It's why hospitals today are (or should be) temples of cleanliness. It's also why doctors and nurses wear facemasks and gloves during surgery, because they might infect their patients, even with their own breath. Now, of course, you can sterilize glass and metal, and linens, scrubs and syringes with heat and chemical substances. But the truth is that plastic and single-use plastic items took medicine into a new era of germ-free perfection, where everything is sterilized, used once, and then disposed of (i.e. never used on another patient).

Single-use plastic is a synonym for life in hospitals because it avoids infections. But it isn't just that. Plastic is so incredibly versatile that we have been able to create a number of amazing new tools for our wellbeing. Many prosthetics and re-ablement tools that allow injured and disabled people to stay mobile and active are often made mostly of plastic. Less invasive surgery, such as arthroscopy or angioscopy, relies on plastic. Instead of sawing your chest open to reach your heart and fix an obstructed artery, today surgeons can perform the same task by inserting a little tube in your groin, reaching your heart and opening up your clogged artery guided by a monitor (made of plastic). This mind-blowing procedure will be done by a doctor completely wrapped in plastic, from his head band, suit and gloves all the way down to his shoes. A complete plastic triumph.

These are some examples of stunning life-saving plastic inventions but plastic has changed our lives for the better in many other ways, too.

PLASTIC TRAVEL

Who didn't push at least one plastic car along the carpet as a kid? Well, now we're driving cars that are often made from lightweight plastics. The most fuel-efficient cars need lightweight build systems and plastic fits the bill. It's estimated that for every 10% reduction in vehicle weight, you can expect around a 7% reduction in the amount of fuel you use.[54] With current concerns about emissions and the air we breathe, the car industry is looking to plastic to increase fuel efficiency. The ancestor of all cars, Ford's Model T, a hundred years ago was essentially a tonne of steel, wood and leather with a massive 3,000 cc engine that cruised at some 12–18 mph. The equivalent (in terms of performance) car today would weigh just a quarter of the Model T and have a 90% smaller engine.

With car windows and interiors already being made from plastics and composites, car makers are now looking at ways to use the same materials in the body panelling. It's also here, in the car, that plastic can save lives: airbags, shock-absorbent foams and panels have drastically reduced the gravity of crash injuries.

By 2020, the average car will incorporate nearly 350kg of plastics, up from 200kg in 2014, according to analyst IHS Chemical (Englewood CO), and these plastics will allow for not just wholesale structural changes but also completely new vehicle designs and concepts. CEO of the Society of Plastics Engineers, Willem De Vos, says that in the 1950s car interiors were about 50% wood and leather; now they're pretty much 100% plastic.[55] On the outside, plastic is used in bumpers and mirrors, and leading manufacturers such as BMW are already looking at the first carbon fibre-filled plastic material for body panels. De Vos suggests glass-reinforced plastic wheels are not far off and plastic used in the exhaust and transmission systems could be commonplace. The plastic car is no longer a concept of dreams.

Lightweight materials are even more important for air travel. Fibre-reinforced plastics in the Boeing 787 Dreamliner resulted in fuel efficiency similar to that of a family car (when measured by kilometres travelled per person). And carbon fibre, the aerospace fibre of choice, is produced from – that's right – plastic.[56]

The wing boxes of the Airbus A380 use plastic fibre composites which reduces the weight of the aircraft by 1.5 tonnes . . . that means it can fly further and carry more cargo using the same amount of fuel.

On the tracks, the rail industry is already using its fair share of recycled plastics as opposed to new wood or plastic or recycled wood. Network Rail is already replacing wooden sleepers with recycled plastic ones. Plastic has the edge because it doesn't rot in bad weather. Recycled plastic wheel chocks and holders for tools that fit on to rail lines during maintenance are already widely used and the industry recognizes that there are other areas where recycled plastics could be explored.

Weight saving is as vital in rail travel as it is in the air and on the road. The less trains weigh, the more economical they are to run. Thermoplastic materials are being considered for the manufacturing of railway carriages because of the many possibilities for weight reduction they offer, without limiting design and durability.

PLASTIC ENERGY

We're always looking for more cost-effective ways to heat our homes and increasing numbers of people are going for solar panels. These roof-top additions to the family home are largely made from plastic!

OCEAN-COMPATIBLE SPORTS GEAR

In 2016 adidas released a good-looking trainer with uppers made using plastic recovered from the ocean. In the same year they unveiled environmentally friendly football shirts for Bayern Munich and Real Madrid as part of its partnership with ocean conservation group and collaboration network Parley. The new shirts are created from Ocean Plastic®, an eco-innovative material created by Parley from upcycled marine plastic waste intercepted from shorelines and in coastal communities. They feature the club logo, three stripes and sponsors' logos in the same colour as the kit. At the time Bayern Munich midfielder, Xabi Alonso said:

'It's an honour to go on this journey with adidas and to have the opportunity to wear the adidas x Parley kit on-pitch for the first time. With every second breath we take coming from our oceans, it's really important that we do what we can do to safeguard them. Wearing a kit that is made from recyclable ocean waste is something I'm very happy about as it's a fantastic opportunity to raise awareness about the need to protect and preserve our oceans. I know this is the start of something very special.'

Real Madrid defender Marcelo, said:

'The ocean is a place I hold close to my heart after growing up in Rio de Janeiro and I have fond memories of playing on the beach when I was a kid. It's amazing to be part of this project and to know that the club I love is making a difference in helping to keep the oceans clean.'[57]

Plastics also help make sports safety gear – such as plastic helmets, mouth guards, goggles and protective padding – lighter and stronger to keep sports enthusiasts of all ages safe. Moulded, shock-absorbent plastic foam helps keep feet stable and supported, while rugged plastic shells covering helmets and pads help protect heads, joints and bones.

5 THINGS THAT HAVE ALREADY BEEN DONE TO BENEFIT OUR LIVES WITH PLASTIC

1. Many prosthetics help disabled people in their daily lives and are often made mostly of plastic.

2. Potentially life-saving surgery, such as arthroscopy or angioscopy, relies on plastic.

3. Cars, trains and planes all use plastic to become safer, more efficient and cheaper.

4. Solar panels for renewable energy are made from plastic.

5. Trainers and sports kit can be made from recycled plastic. Sports safety gear such as helmets and goggles are stronger and safer to keep us protected.

FAREWELL

Our journey through a de-plasticized day ends here. Your life might not be completely detoxed from plastic yet, but your knowledge and awareness make you a different person already. If this 'experience' did what it set out to do, we should have achieved two objectives. Firstly, you now notice 'it', plastic, all around you (literally). Secondly, once you've noticed, you ask yourself: is this necessary? Where does it come from? Where will it end?

The answers to these questions will make you a different person because you'll realize that, no, wrapping two courgettes in plastic is not necessary (and you will not buy them), and the same goes for the plastic object that does not say 'made from recycled plastic'. And when you're left with a plastic item at the end of its life you will never, ever, chuck it in the general waste bin. A detox was never this easy.

Our book has focused on what we can do as individuals. The truth is that we cannot do it alone, we need better laws and better businesses. But don't despair, there's a lot we can do to influence businesses and institutions. Institutions of all shapes and sizes must listen, at some point, to our demands. Whether it's the local council and the colour and/or number of your bins, or your local MP when it comes to a new law on the way we produce things. You have a say.

Search for information and you'll probably discover that there is already a local association that is trying to convince your mayor, there's already a petition that addresses your government or parliament. Join them, sign up, support people that share your concerns.

When it comes to business, it's easy: just vote with your wallet.

We do not need an ideological war against plastic. The worst we could do is to fall into the trap of a stereotype: contemporary/artificial is bad, old (wood and iron) is good.

Progress is good, cities are good, new materials can be amazing, especially plastics. This is not about old versus new, traditional and home-made versus industrial and globalized. This is about life. The reason why we need to change the way we produce and consume plastic is because our current system is destroying life. And that is simply wrong.

We fell in love, a long time ago, with life in the sea. And the sea is the place where our plastic overdose shows its most lethal effects.

So, once you've 'detoxed', head to the coast, look at the horizon and listen to the words the wind is carrying: T H A N K Y O U. It is life, underwater, thanking you.

Here, above water, these words simply come from us: Clare, Luca, Richard and the entire crew at the Marine Conservation Society.

Thank you.

APPENDIX

A BRIEF HISTORY OF PLASTIC AND RECYCLING

Coal and oil have fuelled our lives since before the Industrial Revolution, right through to the modern day. We know now that the air we breathe is tainted with particles that come from these fossil fuels; they damage our lungs, even if they are too small to see. We're aware that some of the gases released in burning fuels are causing Earth's atmospheric temperature to rise quickly, and even affecting the chemistry of the oceans. All of this is due to the enormous scale of fossil-fuel use. And it is the availability of oil, in particular, and its amazing properties, that has given the world plastic.

Oil is simply a build-up of prehistoric marine animals – billions of tiny plankton and algae – which died 300 million years or so ago. Locked away from oxygen, they decayed under pressure beneath the seabed to form a thick liquid substance that's rich in carbon and hydrogen. This liquid can be refined into separate, remarkably consistent fractions that supply the fuel and raw materials for modern industry.

Plastics are made from the hydrocarbons in oil. While materials with 'plastic' qualities exist in nature (cellulose in plant cells that help trees stand up tall and strong, for example), oil-derived plastics are cheap to create into an assortment of products, consistently, time and time again.

One of the first plastics to come into the modern home was Bakelite, made commercially by Leo Hendrick Baekeland in 1910 from phenol (extracted from coal tar) and formaldehyde (made from methanol). You could switch the light on with a Bakelite switch, walk on Bakelite floor tiles, and call distant relatives on a Bakelite phone. It needed some rather toxic chemicals to make it, and asbestos was sometimes mixed in with the plastic. Bakelite has largely been consigned to history now.

New discoveries followed, like polyvinyl acetate in 1913 and polystyrene in 1929, and the 1930s gave rise to a sudden rush of new plastics that are still very much in use today. Nylon was a big break-through in the 1930s, giving rise to a rush for silk-like stockings as well as parts for newly mass-produced cars. It helped make the DuPont company, originally a maker of gunpowder and dynamite, one of the biggest corporations in the world.

Plastics played a big part in the Second World War, with nylon becoming the material of choice for parachutes, Perspex and Bakelite used in aircraft instruments, and new plastics like silicones and polyethylene terephthalate (or PET, since used in billions of plastic bottles) developed.

The investment in factories for wartime production led to plastics being readily available in the peacetime that followed. Tupperware for the home, Teflon for non-stick cooking pans, low-density polyethylene for squeezy toothpaste tubes . . . Global production of plastics really took off in the 1950s, bringing plastic into the lives of a generation of people who could make full use of it, and who began to take it for granted, too.

There are now thousands of different plastic polymers in use, made to serve purposes like containing ready meals at oven temperatures, making a barrier to ultra-violet light for sunglass lenses, and scratch-resistant visors on astronaut helmets. The different chemicals and treatments involved mean that only some of these varied materials are readily recycled.

Packaging makes us buy more things. What's the difference between one type of spring water and another (or tap water, too!)? Really, just the bottle it is presented in, and the label. Unfortunately, the techniques used to make that bottle or chocolate bar wrapper stand out from the next one – its shape, colour and brand – can make the material that much harder to recycle after it has been used just once.

Table of recyclable material types

These are the codes used widely on plastic items available in the UK and worldwide.

CODE	NAME	USES	RECYCLING
△ 1 PET	Polyethylene terephthalate	Drinks bottles, clear food trays and punnets, polyester fabrics.	Widely collected and recycled throughout the UK.
△ 2 HDPE	High density polyethylene	Milk bottles, thin carrier bags, shampoo and detergent bottles and drainpipes.	Widely collected and recycled throughout the UK. Can be made into new milk bottles, and pipes.
△ 3 PVC	Polyvinyl chloride	Window frames, toys, medical bags and tubing, wiring insulation.	Some companies can recycle PVC to use in new products, but it is not widely collected for recycling as it is more expensive to do so than make new.

♷ **LDPE**	Low density polyethylene	Thicker carrier and re-use bags, four or six-pack yokes holding cans together, lining coffee cups and other paper/card containers.	Only a small number of companies can recycle LDPE. It is not very widely collected for recycling. New products are generally of a lower grade than the original material e.g. black bin bags.
♸ **PP**	Polypropylene	Bottle tops. Ropes and fishing nets. Carpets and rugs. Plant pots. Sweet wrappers.	Not widely collected for recycling although it can be made into new products.
♹ **PS**	Polystyrene	Takeaway cartons, coffee cups, protective packaging.	Not widely collected for recycling although it can be made into new products.
♺ **OTHER**	Other	Includes nylon, blended plastics, and most other types of plastic in use today.	Not collected for recycling.

Clean plastic has a value – it isn't really waste when you can turn it into a new product, especially when it is cheaper to do that than to make it new from oil. A plastic bottle, for example, takes 75% less energy to make from recovered bottles than to make new, according to the British Plastics Federation[58] so it should make financial sense to recycle.

But it isn't quite that simple. Oil prices fluctuate enormously, and have been relatively low for a number of years up until 2018. This makes the price offered per tonne on those waste PET plastic bottles lower than it might otherwise be – and, for local authorities and companies dealing with waste, it means less income.

What we do in our daily lives can make a difference. If you'd like to see less plastic in use, and more and better recycling please go to mcsuk.org/appeal/stop-fast-food-plastic and stop the plastic tide.

REFERENCES

EPIGRAPH

Carson, Rachel, 1951. *The Sea Around Us*. Oxford: Oxford University Press.

Attenborough, Sir David. 2017. *Blue Planet II*, BBC One.

INTRODUCTION

1. Andrés, C., Martí, E., Duarte, C.M., García-de-Lomas, J., van Sebille, E. *et al*, 2017. The Arctic Ocean as a dead end for floating plastics in the North Atlantic branch of the Thermohaline Circulation. *Science Advances* 19 Apr: 3(4), e1600582, DOI: 10.1126/sciadv.1600582.

2. Wiig O., Derocher A.E., Cronin M. and Skaare J.U., 1998. Female pseudohermaphrodite polar bears at Svalbard. *Journal of Wildlife Diseases* 34(4), pp.792–796.

3. Reijnders, P. J. H., 1982. On the ecology of the harbour seal *Phoca vitulina* in the Wadden Sea: Population dynamics, residue levels, and management. *Veterinary Quarterly* 4(1), pp. 36–42.

4. Gilmartin, W.G, Delong, R.L., Smith, A.W., Sweeney, J.C. *et al*, 1976. Premature Parturition in the California Sea Lion. *Journal of Wildlife Diseases*, 12(1):104–115.

5. Van Franeker, J.A., Blaize, C., Danielsen, J., Fairclough, K., *et al*, 2011. Monitoring Plastic Ingestion by the Northern Fulmar *Fulmarus glacialis* in the North Sea. Environmental Pollution 159, 2609–2615.

6. World Economic Forum, Ellen MacArthur Foundation and McKinsey & Company, 2016. The New Plastics Economy: Rethinking the future of plastics. Available at: http://www.ellenmacarthurfoundation.org/publications.

7. Marine Conservation Society, 2017. Great British Beach Clean 2017 Report. Available at: https://www.mcsuk.org/media/GBBC_2017_Report.pdf [accessed 29th April 2018].

8. A Speech by HRH The Prince of Wales for the Our Ocean Conference, Malta, 5 October 2017. Available at: https://www.princeofwales.gov.uk/speech/speech-hrh-prince-wales-our-ocean-conference-malta [accessed 12th May 2018].

GETTING UP

9. Rhodes, C., 2018. Manufacturing: statistics and policy. House of Commons Library Briefing Paper Number 01942. 2 January. Available at: http://researchbriefings. parliament.uk/ResearchBriefing/Summary/ SN01942#fullreport [accessed 29th April 2018].

10. Baldé, C.P., Forti V., Gray, V., Kuehr, R., Stegmann,P., 2017. The Global E-waste Monitor. United Nations University (UNU), International Telecommunication Union (ITU) & International Solid Waste Association (ISWA), Bonn/Geneva/Vienna. Available at: https:// collections.unu.edu/eserv/UNU:6341/Global-E-waste_ Monitor_2017__electronic_single_pages_.pdf [accessed 29th April 2018].

11. British Broadcasting Corporation, 2011. Track My Trash, *Panorama*, BBC One, 16 May. Available at: https:// www.bbc.co.uk/programmes/b0116gw0 [accessed 29th April 2018].

12. Waste & Resources Action Programme, 2011. WRAP Information Sheet: The value of re-using household WEEE. Available at: http://www.wrap.org.uk/sites/files/ wrap/WRAP%20WEEE%20HWRC%20summary%20 Sheet.pdf [accessed 29th April 2018].

European Commission (2018) http://ec.europa.eu/ eurostat/statistics-explained/index.php/Waste_

statistics_-_electrical_and_electronic_equipment
[accessed 15th May 2018].

BATHROOM

13. Wright, L., 1960. *Clean and decent: the fascinating history of the bathroom & the water closet and of sundry habits, fashions & accessories of the toilet principally in Great Britain, France, & America*. Routledge & Kegan Paul p. 2.

14. Women's Environmental Network, 2018. https://www.wen.org.uk/environmenstrual/ [accessed 12th May 2018].

BABIES AND TODDLERS

15. Mughal, B.B., Fini, J.B., and Demeneix, B.A., 2018. Thyroid-disrupting chemicals and brain development: an update. *Endocrine Connections* 7:R160-R186; published ahead of print March 23, 2018, doi:10.1530/EC-18-0029.

16. Tsai, M-J., Kuo, P-L., and Ko, Y-C., 2012. The association between phthalate exposure and asthma. The Kaohsiung Journal of Medical Sciences. Volume 28, Issue 7, Supplement, Pages S28-S36 https://doi.org/10.1016/j.kjms.2012.05.007.

17. Wolff, M. S., Teitelbaum, S. L., McGovern, K., Windham, G. C., Pinney, S. M., Galvez, M., Calafat, A.M., Kushi, L.H., and Biro, F.M. on behalf of the Breast Cancer and Environment Research Program. (2014). Phthalate

exposure and pubertal development in a longitudinal study of US girls. *Human Reproduction*, 29(7), 1558–1566. http://doi.org/10.1093/humrep/deu081.

Kim, S. H., & Park, M. J., 2014. Phthalate exposure and childhood obesity. Annals of Pediatric Endocrinology & Metabolism, 19(2), 69–75. http://doi.org/10.6065/ apem.2014.19.2.69.

18. https://www.huffingtonpost.co.uk/entry/nhs-purchases-staggering-half-a-billion-disposable-cups-every-5-years_uk_5accc512e4b0e37659b0d8f2 [accessed 12th May 2018].

19. Turner, A., 2018. Concentrations and Migratabilities of Hazardous Elements in Second-Hand Children's Plastic toys. *Environmental Science & Technology* 52 (5), 3110-3116 DOI: 10.1021/acs.est.7b04685.

20. Environment Agency, 2008. An updated lifecycle assessment for disposable and reusable nappies. Science Summary SC010018/SS2. Available at: https://www.gov. uk/government/publications/an-updated-lifecycle-assessment-for-disposable-and-reusable-nappies [accessed 29th April 2018].

21. Kramer, M.S. and Kakuma, R., 2002. The Optimal Duration of Exclusive Breastfeeding: A Systematic Review. World Health Organisation, Department of Nutrition for Health and Development. Cochrane Database of Systematic Reviews, Issue 1. Art. No.: CD003517. DOI: 10.1002/14651858.

22. Centers for Disease Control and Prevention, 2017. Breastfeeding Rates, National Immunization Survey. Available at: https://www.cdc.gov/breastfeeding/data/nis_data/index.htm [accessed 29th April 2018].

23. DiGangi, J., Strakova, J. and Bell, L., 2017. POPS Recycling Contaminates Children's Toys with Toxic Flame Retardents. *Arnika and IPEN*. Available at: http://ipen.org/sites/default/files/documents/toxic_toy_report_2017_update_v2_1-en.pdf [accessed 29th April 2018].

24. Turner, A., 2018. Concentrations and Migratabilities of Hazardous Elements in Second-Hand Children's Plastic toys. Environmental Science & Technology. 52, 5, 3110-3116 doi: 10.1021/acs.est.7b04685.

25. IPEN, 2013. Available at: http://www.ipen.org/news/provisional-agreement-ban-hbcd [accessed 12th May 2018].

26. Babycentre, 2018. Pushchair safety: what you need to know. Available at: https://www.babycentre.co.uk/a559727/pushchair-safety-what-you-need-to-know#ixzz5Dz24X2mp/ [accessed 12th May 2018].

THE OFFICE

27. United Nations Department of Economic and Social Affairs, Population Division. 'World Urbanisation Prospects, the 2014 Revision'. The United Nations, ISBN

978-92-1-151517-6. Available at: https://esa.un.org/unpd/wup/publications/files/wup2014-highlights.pdf [accessed 29th April 2018].

28. Breitburg, D., Levin, L.A., Oschlies, A., Grégoire, M., Chavez, F.P., Conley, D.J., Garçon, V., Gilbert, D., Gutiérrez, D., Isensee, K., Jacinto, G.S., Limburg, K.E., Montes, I., Naqvi, S.W.A., Pitcher, G.C., Rabalais, N.N., Roman, M.R., Rose, K.A., Seibel, B.A., Telszewski, M., Yasuhara, M., and Zhang, J., 2018. Declining oxygen in the global ocean and coastal waters. *Science*. Jan 5;359(6371). P. ii: eaam7240. doi: 10.1126/science. aam7240.

29. Euromonitor International 2017 Packaging edition, reported *Guardian*, 2017. A million bottles a minute: world's plastic binge 'as dangerous as climate change'. *Guardian*, 28 June. https://www.theguardian.com/environment/2017/jun/28/a-million-a-minute-worlds-plastic-bottle-binge-as-dangerous-as-climate-change' [accessed 29th April 2018].

AFTER-WORK ACTIVITIES

30. State of the Nation Fitness Industry reported in http://www.leisuredb.com/blog/2017/5/5/2017-state-of-the-uk-fitness-industry-report-out-now [accessed 12th May 2018].

31. Duggal N.A., Pollock R.D., Lazarus N.R., Harridge S. and Lord J.M., 2018. Major features of immunesenescence,

including reduced thymic output, are ameliorated by high levels of physical activity in adulthood. *Aging Cell.* 17(2). doi: 10.1111/acel.12750. Epub 2018 Mar 8.

32. 42.3 billion value comes from: Eventbrite, 2016. An Introduction to Events; The UK Events Industry in numbers. Available at: https://www.eventbrite.co.uk/blog/uk-event-industry-in-numbers-ds00/ [accessed 29th April 2018].

33. Rugby Football Union, 2018. Available at: https://www.twickenhamexperience.com/fan-cup [accessed 12th May 2018].

PETS

34. Petopedia, Dog Poo & You. Available at: https://petopedia.petscorner.co.uk/dog-poo-you/ [accessed 14th May 2018].

    Statista, 2018. Estimated pet population in the United Kingdom (UK) from 2009 to 2018. Available at: https://www.statista.com/statistics/308229/estimated-pet-population-in-the-united-kingdom-uk/ [accessed 14th May 2018].

35. Totally Money, 2016. Furry Finances: The Average Cost of Raising a Pet is £10,975. Totally Money, Consumer

News, 20th June. Available at: http://www.totallymoney.
com/press-centre/furry-finances-average-cost-raising-pet-
10975/ [accessed 29th April 2018].

36. Animal Friends Insurance, 2018. Available at: https://
www.animalfriends.co.uk/blog/are-you-aware-of-the-new-
dog-poo-laws/ [accessed 12th May 2018].

37. ibid.

WASTE OF TIME

38. CB Environmental, unknown date. 20 Facts About Waste
and Recycling. Available at: https://www.
cbenvironmental.co.uk/docs/Recycling%20Activity%20
Pack%20v2%20.pdf [accessed 29th April 2018].

39. ibid.

40. Defra and Government Statistical Service, 2018. UK
Statistics on Waste. 22nd February. Available at: https://
assets.publishing.service.gov.uk/government/uploads/
system/uploads/attachment_data/file/683051/UK_
Statisticson_Waste_statistical_notice_Feb_2018_FINAL.
pdf [accessed 29th April 2018].

41. ibid.

42. ibid.

43. REcoup., 2017. UK Household Plastics Collection Survey 2017. Available at: http://www.recoup.org/p/229/uk-household-plastics-collection-survey-2016/ [accessed 12th May 2018].

44. ibid.

THE WEEKEND

45. Aldridge, S., and Miller, L., 2012. *Why Shrink-wrap a Cucumber? The Complete Guide to Environmental Packaging.* London: Lawrence King Publishing.

46. Dairy UK., 2018. http://www.dairyuk.org/ industry-overview/consumption-sales [accessed 12th May 2018].

47. *Daily Mail,* 2007. http://www.dailymail.co.uk/femail/ article-469180/Why-fallen-stocking-sales.html [accessed 12th May 2018].

48. *Independent,* 2007. https://www.independent.co.uk/news/uk/ this-britain/a-sales-shock-a-last-glimpse-of-stockings-5334061.html [accessed 12th May 2018].

49. Napper, I.E. and Thompson, R., 2016. Release of Synthetic Microplastic Plastic Fibres From Domestic Washing Machines: Effects of Fabric Type and Washing Conditions. *Marine Pollution Bulletin,* Volume 112, Issues 1–2, 15. Pages 39–45.

50. Bloomberg, with assistance by Shuping Niu and Ken Wills, 2017. Plastic Film Covering 12% of China's Farmland Pollutes Soil. *Bloomberg* [online] 5th September, revised 6th September. Available at: https://www. bloomberg.com/news/articles/2017-09-05/plastic-film- covering-12-of-china-s-farmland-contaminates-soil [accessed 29th April 2018].

SPECIAL OCCASIONS

51. Cadbury World. Date unknown. Cadbury's Easter Brand Factsheet. Available at: https://www.cadburyworld.co.uk/ schoolandgroups/~/media/CadburyWorld/en/Files/Pdf/ factsheet-easterbrands [accessed 29th April 2018].

52. Crowe, V., 2018. The Truth About Easter Egg Packaging. *Which?* Available at: https://www.which.co.uk/news/2018/03/ the-truth-about-easter-egg-packaging/ [accessed 29th April 2018].

PLASTIC SAVES LIVES

53. Clark P.F. (1920). Joseph Lister, his Life and Work. *The Scientific Monthly* Vol. 11, No. 6, pp. 518–539; published by: American Association for the Advancement of Science, Stable URL: http://www.jstor.org/stable/6707.

54. IHS Chemical (Englewood CO), 2015, reported in Plastics Today. Plastics use in vehicles to grow 75% by 2020, says industry watcher. Available at: https://www.

plasticstoday.com/automotive-and-mobility/plastics-use-vehicles-grow-75-2020-says-industry-watcher/
63791493722019 [accessed 30th April 2018].

55. Pickering, K., 2017. The future of plastics: reusing the bad and encouraging the good. The Conversation.com, November 20. Available at: http://theconversation.com/the-future-of-plastics-reusing-the-bad-and-encouraging-the-good-87001 [accessed 30th April 2018].

56. Ibid.

57. adidas press release for the launch of 'adidas x Parley for the Oceans Real Madrid and Bayern Munich home jerseys' campaign. November 2016. Available at: https://news.adidas.com/GLOBAL/Latest-News/adidas-and--parley-for-the-oceans-release-real-madrid-and-bayern-munich-jerseys-made-from-parley-oce/s/dd05b722-d34c-41f9-82ac-0c7d0e34f262

APPENDIX

58. Publications. Parliament. UK. First Report of Session 2017–19, on Plastic bottles: Turning Back the Plastic Tide, HC 339 on 22 December 2017. Available at: https://publications.parliament.uk/pa/cm201719/cmselect/cmenvaud/841/84102.htm. [Accessed 12th May 2018].

FURTHER READING

Murray F. and Cowie P.R., 2011. Plastic contamination in the decapod crustacean *Nephrops norvegicus* (Linnaeus, 1758). *Marine Pollution Bulletin.* Jun;62(6):1207–17. doi: 10.1016/j. marpolbul.2011.03.032. Epub 2011 Apr 16.

Ward, J.E., Kach, D.J., 2009. Marine Aggregates Facilitate Ingestion of Nanoparticles by Suspension-Feeding Bivalves. *Marine Environmental Research*, doi: 10.1016/j.marenvres.2009.05.002.

Andrady, A., 2011. Microplastics in the marine environment. *Marine Pollution Bulletin*, 62(8), pp.1596–1605.

Engler, R., 2012. The Complex Interaction between Marine Debris and Toxic Chemicals in the Ocean. *Environmental Science & Technology.* 46. 10.1021/es3027105.

Plastics Historical Society http://plastiquarian.com [accessed 29th April 2018] for facts about plastic research and development.

ACKNOWLEDGEMENTS

Thank you to our colleagues at the Marine Conservation Society, and all of the volunteer beach cleaners, letter writers and activists who, over the last quarter of a century and more, have inspired this work.

Thank you in particular to our pollution experts in the Clean Seas and Beachwatch teams, both present and past, and the wider teams around the UK; and to the original plastic challengers Emily Smith and Kate Wilson.

Thanks to our partners in the fight against grime, including Campaign for the Protection of Rural England, Association for the Protection of Rural Scotland, Fauna and Flora International, Greenpeace, Two Minute Beach Clean, Surfers Against Sewage, Ocean Conservancy and many more. Thanks, too, to Waitrose and players of People Postcode Lottery, for funding our current beach clean work.

Thank you to Sir David Attenborough, for a life spent reminding us all of the importance and beauty of nature.

And for taking the message to a whole new level in 2017 with *Blue Planet II*.

Last but not least, thank you to our families and friends who we have neglected during the writing of this book.

ANNUAL MARINE CONSERVATION SOCIETY EVENTS

March: World Water Day

April: Plastic-Free Easter

May: International Plastic Bag-Free Day

June: World Oceans Day; World Environment Day

July: Plastic Challenge Month

September: Great British Beach Clean

October: World Habitat Day

November: World Toilet Day

December: Plastic-Free Christmas